Lecture Notes in Electrical Engineering

Volume 104

Unai Alvarado, Guillermo Bistué,
and Iñigo Adín

Low Power RF Circuit Design in Standard CMOS Technology

Springer

Unai Alvarado
CEIT and TECNUN
(University of Navarra)
Paseo Mikeletegi, Parque
Tecnológico Miramon 48
20009 San Sebastián
Spain
Email: ualvarado@ceit.es

Iñigo Adín
CEIT and TECNUN
(University of Navarra)
Paseo Mikeletegi, Parque
Tecnológico Miramon 48
20009 San Sebastián
Spain
Email: iadin@ceit.es

Guillermo Bistué
CEIT and TECNUN
(University of Navarra)
Paseo Mikeletegi, Parque
Tecnológico Miramon 48
20009 San Sebastián
Spain
Email: gbistue@ceit.es

ISBN 978-3-642-22986-2 e-ISBN 978-3-642-22987-9

DOI 10.1007/978-3-642-22987-9

Lecture Notes in Electrical Engineering ISSN 1876-1100

Library of Congress Control Number: 2011934148

Typeset & Cover Design: Scientific Publishing Services Pvt. Ltd., Chennai, India.

Printed on acid-free paper

9 8 7 6 5 4 3 2 1

springer.com

Este libro está dedicado a nuestras familias,
por su paciencia y apoyo incondicional
Unai, Guillermo e Iñigo

Acknowledgements

We would like to acknowledge both the Consejería de Educación, Universidades e Investigación of the Basque Government and the Ministerio de Innovación y Ciencia of the Spanish Government for their continuous support to our research group.

We would also like to express our gratitude to all the staff at CEIT and TECNUN (University of Navarra) and our colleagues at University of Sevilla and IUMA Research Centre, for the fruitful collaboration throughout the last years.

Finally, we would like to mention the people who helped us in the review of the manuscript, and the staff of Springer DE for their support, especially Mayra Castro, Petra Jantzen and Suguna R.

Contents

List of Figures

List of Tables

1 Introduction

One of the main demands of electronic devices users has been mobility. Not so long ago, this kind of equipment was necessarily associated with an electric cord and a power outlet. In the best-case scenario, energy was supplied by some rather inefficient chemical batteries. However, during the last years this situation has changed dramatically. On the one hand, the development of low cost and high performance rechargeable batteries has allowed a certain degree of autonomy. On the other hand, a great effort has been dedicated to reduce the power consumption needs.

In the next sections we present a brief report of the current situation, starting with the recent evolution of the modern handheld devices.

1.1 Recent Evolution of Personal Communication Devices

History tells that wireless communications have existed for more than a century. The original application was invented in July 1896 over a distance of one-and-three-fourths miles on Salisbury Plain. This is where Guglielmo Marconi demonstrated his efforts at creating an operational wireless telegraph apparatus, to the British Telegraph authorities. Records show his first British patent application was filed on June 2 of that year. From 1890 until 1896 many American science students held a keen interest and studied discoveries made in Europe; but it was not until 1897 that the utilitarian American mind sensed the commercial possibilities of advances being made abroad. From that moment in engineering history, people have continued to work on wireless systems, in a bid to increase the distance and the utility.

The idea of the first cellular network was brainstormed in 1947. It was intended to be used for military purposes as a way of supplying troops with more advanced forms of communications. From 1947 until circa 1979 several different forms of broadcasting technology have emerged. The United States began to develop the AMPS (Advanced Mobile Phone Service) network. The frequencies allocated to AMPS by the FCC ranged between 824 to 849 MHz in reverse channels (mobile to base) and 869 to 894 MHz in forward channels (base to mobile).

Europeans quickly realized the disadvantages of each country operating on their mobile network. It prevents cell phone use from country to country within Europe. In 1982, with the emerging European Union and high travel volume between countries in Europe, the Conference of European Posts and Telegraphs (CEPT) assembled a research group with intentions of researching the mobile phone system in Europe. This group was called Group Spéciale Mobile (GSM).

U. Alvarado et al.: Low Power RF Circuit Design in CMOS Tech., LNEE 104, pp. 1–9.
springerlink.com

For the next ten years the GSM group outlined standards, researched technology and designed a way to implement a pan-European mobile phone network. In 1989 work done by the GSM group was transferred to the European Telecommunication Standards Institute (ETSI). The name GSM was transposed to name the type of service invented. The acronym GSM had been changed from Group Spéciale Mobile to Global Systems Mobile Telecommunications. By April of 1991 commercial service of the GSM network had begun. Just a year and half later in 1993 there were already 36 GSM networks in over 22 countries. Several other countries wanted to adopt this new mobile phone network and participate in what was becoming a worldwide standard. Nowadays, GSM-900 and GSM-1800 are used in most parts of the world: GSM-900 uses 890 - 915 MHz to send information from the Mobile Station to the Base Transceiver Station and 935 - 960 MHz for the other direction. Moreover, GSM-1800 uses 1710 - 1785 MHz to send information from the Mobile Station to the Base Transceiver Station and 1805 - 1880 MHz for the other direction. At present, in North America, GSM operates on the primary mobile communication bands GSM-850 MHz and GSM-1900 MHz.

Mass wireless applications did not initially emerge and only came with the advent of the personal cellular phones and the need of data transmission. Emergency services such as the police department utilize wireless networks to communicate important information quickly. People and businesses use wireless networks to send and share data quickly whether it is in a small office building or across the world. The data transmission necessities come from what it is nowadays call Internet. Historically, the first data network was established between the University of California, Los Angeles and the Stanford Research Institute, on 29 November 1969. This was known as ARPANET. By 5 December 1969, a 4-node network was connected by adding the University of Utah and the University of California, Santa Barbara. Building on ideas developed in ALOHAnet, the ARPANET started in 1972 and was growing rapidly, even with wireless satellite links. By 1981 the number of hosts had grown to 213, with a new host being added approximately every twenty days. Internet means a global and large network using the TCP/IP protocols (Transmission Control Protocol / Internet Protocol). As interest in wide spread networking grew and new applications for it arrived, the Internet's technologies spread throughout the rest of the world. Wireless connections to Internet offer, besides, more mobility.

1.2 Examples of Applications

In the first decade of 21st century the number of mass wireless communications devices has grown exponentially. Among the reasons of this spectacular spread we could note the definition of several dedicated standards, targeting both professional and consume markets. Fig. 1.1lists some examples.

Fig. 1.1 World Wide Wireless applications

These common wide spread applications are based on standards such as shown in Fig. 1.1. This collection of logos illustrates the enormous variety of associations implied in the wireless technology world, and its significance in the society. Furthermore, each standard has its own purpose. Some are focused on low data rate close connections, other need to push their data rate to the maximum to assure a higher speed in the communications, other's goal is an intermediate throughput but with higher security.

Those standards have been adopted by many communications platforms, most of them working in the up-to-5 GHz band. Examples are listed below:

- Cellular phones and pagers: provide connectivity for portable and mobile applications, both personal and business. GSM-800, GSM-850, GSM-1800 and GSM-1900 have been the most widespread services. And in the coming years 3G and LTE services will hand over the growing demand for mobile communications.
- Global Positioning System, GPS: allows drivers of cars and trucks, captains of boats and ships, and pilots of aircraft to ascertain their location anywhere on earth. Nowadays, the European GALILEO is being launched. Both systems are transmitted in two main signals L1 and L2, with central frequency bands at 1575 MHz and 1227 MHz.

- Cordless computer peripherals: the cordless mouse is a common example; keyboards and printers can also be linked to a computer via wireless. Nowadays, the most common standard for this application is Bluetooth, which carrier is at 2400 MHz.
- Cordless telephone sets: these are limited-range devices, not to be confused with cell phones. The frequency allocations of these devices has evolved in parallel with the technology: 900 MHz (902–928 MHz) allocated in 1990, 1.9 GHz (1920-1930 MHz) developed in 1993 and allocated U.S. in October 2005 and 2.4 GHz allocated in 1998.
- Wireless Sensors Networks: With these networks it is possible to achieve large scale continuous environmental monitoring, using hundreds of operation nodes. Other applications include biometric parameters measurement or large structure health assessment.
- DVB-H standard was adopted in 2004 by ETSI as the system specification for broadcasting mobile DTV services to handheld devices. DVB-H is based on DVB-T, but it introduces new features that guarantee good quality of service while assuring indoor coverage, power consumption saving, transmission of multimedia contents, and better mobility.
- Radio Frequency Identification (RFID) and passive sensor associated technologies: RFID was conceived as a substitute for barcode identifiers. However, the combination of this technology with low power sensors (MEMs and CMOS mainly) has allowed the development of passive sensor nodes. RFID communications operate both in the short (inductive) and far field, with frequencies up to 2.5 GHz.

The common factor among all wireless applications lies in their ability to provide dynamic real-time services free of interferences. These two conditions are based on the frequency allocation, the available bandwidth for the channels and the modulations of the standards. On top of this, there is a permanent driving force towards miniaturization and mobility. These conditions imply necessarily low power consumption devices.

1.3 Frequency Allocation for the Next Wireless Applications

Current market demands are pushing wireless communication systems across all environments and all application fields. The unlicensed Industrial Scientific and Medical (ISM) 2.4 GHz band allocates an increasing number of standards, such as Bluetooth and IEEE 802.11b/g. So with the exponential increase in users of these standards, as shown inFig. 1.2, it results in an over population of the available bandwidth in this band.

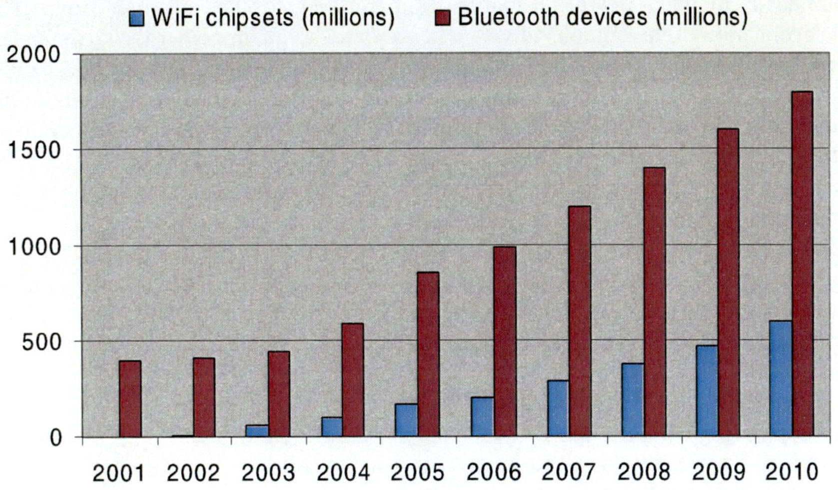

Fig. 1.2 Evolution and prediction of the WiFi users and Bluetooth devices

Consequently, the growth of the modern Broadband Wireless Networks (BWN) has to expand to higher frequency bands, i.e. the Unlicensed National Information Infrastructure (U-NII) 5 GHz band. As described in section 1.2, most of the recently approved or upcoming standards are planned for higher frequencies than 2.4 GHz. IEEE 802.20 focuses the licensed 3.5 GHz, IEEE 802.11a/n and IEEE 802.16a/e which are fixed in the 5 GHz U-NII band and the UWB standards are expanded in 528 MHz wide channels from 3.1 to 10.6 GHz.

Therefore, in the next years will be still necessary to develop more circuits and systems working for frequencies up to 5 GHz.

1.4 Common Requirements to Current Handheld Devices: OFDM Modulations

In parallel with the up migration of the frequency allocation, the efficient spectral utilization becomes more and more important, compensating for the additional complexity involved in the modern broadband wireless standards. This complexity is strengthened by the multistandard conception of the modern wireless electronic devices. Harder system specifications are demanded to meet the standards specifications.

Not that long ago, most forms of electronic communication were analogue from input to output. While this tended to minimize system complexity, the result was a communication link with less than optimal noise rejection, spectral efficiency, and/or reliability. With the rapid advances being made in integrated circuitry, specifically in the field of high-speed digital signal processing, many communication links are now being designed to utilize digital modulation. For a given bandwidth and signal to noise ratio, greater information capacity is attainable by new devices capable to get closer to the Shannon limit. The determination of the broadband wireless standards to attain higher throughputs has led to increased complexity of these digital modulations. Multiple carrier modulation is one possible means of transmitting data symbols in parallel. One motivation for this additional complexity is the frequency diversity inherent in these multiple carriers and their orthogonality. This may be used to overcome channel impairments caused by multipath and other distortions common to RF transmission channels.

The Orthogonal Frequency Division Multiplexing (OFDM) is the basic method of the carrier multiplicity. This technique divides the bandwidth into multiple frequency subcarriers. All these subcarriers are orthogonal to each other and are all modulated with a conventional modulation scheme, such as Quadrature Amplitude Modulation (QAM) at a low symbol rate. The term "OFDM" is frequently followed by the number that depicts the potential number of subcarriers in the signal (including the DC and the guard-band subcarriers), e.g. OFDM-64. The high bit rate digital stream is divided into several low bit rate schemes and is transmitted in parallel.

An evolution of this technique, the Orthogonal Frequency Division Multiple Access (OFDMA) technique, combines OFDM multiplexing method with multiple access using time, frequency or coding separation of the users. The multiplicity of the channel is achieved with the designation of different OFDM subchannels to different users:

- OFDMA employs multiple closely spaced subcarriers
- The subcarriers are divided into groups of subcarriers
- Each group is named a subchannel
- The subcarriers that form a subchannel do not need to be adjacent

In the downlink, a subchannel may be intended for different receivers. In the uplink, a transmitter may be assigned one or more subchannels. This concept of sub-channelization is explained in Fig. 1.3 and is compared to the OFDM basic approach with the same data load.

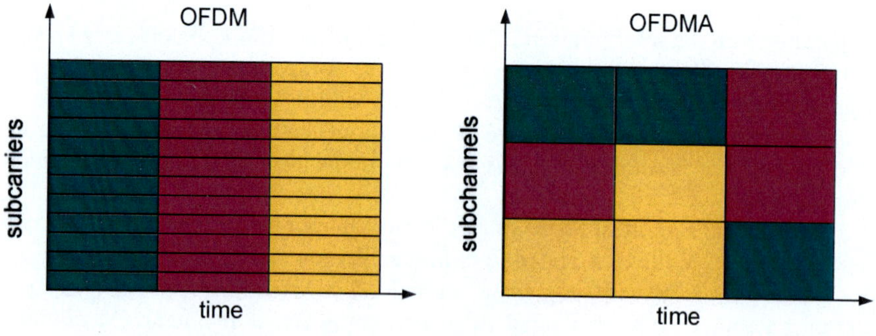

Fig. 1.3 OFDM and OFDMA channel composition

This multiple access multiplexation technique shares all the advantages of the OFDM method applied to the mobile accesses. Essentially, a user on an OFDMA network is assigned a number of sub channels across the band. A user close to the base station would normally be assigned a larger number of channels with a high modulation scheme such as 64 QAM to deliver the most data throughput to that user. As the user moves farther away, the number of subchannels is re-assigned dynamically to fewer and fewer sub channels. However, the power allotted to each channel is raised. The modulation scheme could gradually shift from 16 QAM to Quaternary Phase Shift Keying (QPSK) and even binary phase shift keying (BPSK) at longer ranges. The data throughput drops as the channel capacity and modulation change, but the link maintains its strength.

Multiple Input Multiple Output (MIMO) techniques are known to boost capacity. For high data rate transmission, the multipath characteristic of the environment causes the MIMO channel to be frequency selective. As it has been introduced established, OFDM based methods can transform such a frequency selective MIMO channel into a set of parallel channels, and therefore decrease receiver complexity. The combination of the two powerful techniques, MIMO and OFDM, is very attractive, and has become one of the most promising broadband wireless access schemes.

A MIMO system takes advantage of the spatial diversity obtained by spatially separated antennas in a dense multipath scattering environment. MIMO systems may be implemented in a number of different ways to obtain either a diversity gain to combat signal fading or to obtain a capacity gain.

In any case, these techniques have contributed to increase the capacity of data networks. In terms of circuit design, the impact has been also considerable. The complexity of the digital processing has led to more exigent requirements for the analog building blocks.

1.5 Low Power RFIC Design

Portable and mobile communication terminals, also known as handheld terminals (defined as a light battery powered apparatus), require specific features to the transmission system:

- As battery powered, the transmission system shall offer them the possibility to repeatedly power off some part of the reception chain to increase the battery usage duration.
- As targeting nomadic users, the transmission system shall ease access to the different services when receivers leave a given transmission cell and enter a new one.
- As they are expected to serve various environments of use (indoor and outdoor, pedestrian and inside moving vehicle), the transmission system shall offer sufficient flexibility/scalability to allow reception of services at different speeds, while optimizing transmitter coverage.
- As services are expected to be delivered in an environment suffering high levels of man-made noise, the transmission system shall offer ways of mitigating their effects on the receiving capabilities.

- As modern telecommunication systems aim to provide a generic way to serve handheld terminals, the transmission system shall support different transmission bands and channel bandwidths, co-existing in the same multistandard device.

In a handheld device, one of the most crucial subsystems in terms of performance requirements is the RF front-end. Any reduction in its power consumption leads directly to a longer battery life. Therefore, the selection of the architecture of such a front-end, as well as the minimization of the power consumption of each of its building blocks is key issues for a successful system implementation. However, there is a previous factor that affects dramatically to the viability of the systems, in terms both of performance and cost: the fabrication technology. Today, the best overall results are obtained with submicron CMOS processes, and in the next point a brief discussion about this topic is shown.

1.5.1 CMOS Technology

Up to now, three factors have been critical in the choice of the technology in the competitive RF industry: performance, cost and time to market. However, as it has been shown in the previous paragraphs, due to the increasing importance of mobile communications an extra parameter has to be added to asses the viability technologies: the power consumption.

The processes of fabrication of ICs currently have four main participants. In ascending order of importance considering the volume of fabrication:

- GalliumArsenide (GaAs)
- Silicon Germanium (SiGe) Bipolar and CMOS (BiCMOS)
- Micro Electrical Mechanical Systems (MEMS)
- Complementary Metal Oxide Semiconductor (CMOS):

CMOS processes have attained 80% of the ICs production in the last decade. This level is supported by the immense importance of the digital IC market. Nevertheless, the demand of III-V compound semiconductors (SiGe and GaAs) has fallen more than 50% in this period. Therefore, CMOS is absorbing part of the analogue IC design market. This progressive migration is due to the attractive possibility of integrate the analogue and the digital part of a receiver/transceiver in a unique chip.

AsGa, the precursor of the semiconductor heterostructures, has lost all its power in the IC mass production market. As for power consumption, AsGa requires a higher current for high speed applications than MOS based processes, as its reduced hole mobility strongly limits the performance of its P channel transistors. Only the BiCMOS processes can theoretically compete with this fully integrability of CMOS.

Nevertheless, the preponderance of the CMOS over the BiCMOS processes is based on three main factors: current consumption, chip area and price. First, the CMOS is ideally suited for use in logic applications because of its low current consumption. Furthermore, many of the improvements applied to CMOS fabrication processes cannot be directly transferred to BiCMOS production methods; the simultaneous optimization of the BJT and the CMOS transistors in the same process is not possible. A trade off is needed and also it requires an extra cost. BiCMOS may never offer the low power consumption of CMOS alone. Moreover, in comparison to the BiCMOS, the CMOS processes integrate each transistor in a smaller area. So that, CMOS is undoubtedly the preferred technology for digital ICs, and this instigates the RF designer to improve the high frequency analogue designs in this technology. Within this technology, there is a continued need for further improvements in performances and power consumption.

One of the traditional disadvantages of CMOS technology was the comparatively low transition frequency compared to the BiCMOS alternatives. However, the commercially available .13 or .09 mm processes present transition frequencies well above 100 GHz. Therefore, the broadband wireless standards allocated in the 5 GHz band and intended to be part of low cost and low power devices can be tackled without any operative restriction.

1.5.2 Low Power Design Techniques for Analog Circuits

Firstly, low power design techniques were applied to digital circuits. Most of these were developed during the early nineties, and new solutions are still under investigation. In spite the obvious interest of specific solution for analog systems, until recently it has been difficult to find true implementations. There are two main research lines in the quest for lower power consumption:

- Fabrication technology optimization
- Development of circuit design techniques

This book deals with the second approach. In the next chapters, the different techniques are presented, together with their application to the main building blocks. To better understand such techniques, different circuits examples will be described within chapters 6, 7 and 8. Such examples will be related to DVB-H and WLAN (at UNII band) standards, due to the increase necessity of lower power consumption at higher operation frequencies and bandwidths.

2 Power Considerations in Analog RF CMOS Circuits

This chapter deals with the basic principles of power consumption in RF CMOS analog circuits. These concepts are used extensively throughout the book; therefore the different sections of the chapter are dedicated to the presentation of general definitions and formulas. Section 2.1 introduces the different sources of power dissipation in analogue circuits, regarding both static and dynamic power dissipation mechanisms, from a steady and transient perspective respectively. Section 2.2 reviews the classical considerations for low power digital circuits and their structures. Then, section 2.3 deals with the power supply scaling as a means to achieve low- power benefits, which is highly related to section 2.2, and finally section 2.3 outlines the practical limits in power consumption from a triple perspective: starting from the front-end architecture and CMOS technology constraints through the RF and analog circuits.

2.1 Sources of Power Dissipation

There are four main sources of power dissipation in CMOS circuits:

- Dynamic switching power due to the charging and discharging circuit capacitances (P_{dyn})
- Leakage current power from reverse-biased diodes and subthreshold conduction ($P_{leakage}$)
- Short-circuit current power due to finite signal rise/fall times (P_{short})
- Static biasing power (P_{bias})

The total power consumption in any circuit can be then described by the following equation:

$$P_{total} = P_{dyn} + P_{leakage} + P_{short} + P_{bias} \qquad \text{Eq. 2.1}$$

2.1.1 Dynamic Switching Power

When CMOS circuits switch, the output is either charged up to VDD, or discharged down to GND. In static logic design, the output only transitions if an input transition occurs, while in dynamic logic, the output is pre-charged during half the clock cycle, and transitions can only occur in the second clock phase, depending upon the input values. In both cases, the power dissipated during

U. Alvarado et al.: Low Power RF Circuit Design in CMOS Tech., LNEE 104, pp. 11–24.
springerlink.com

switching is proportional to the capacitive load. The dynamic power dissipation is given by

$$P_{dyn} = \alpha \cdot C_L \cdot VDD^2 \cdot f_s \quad \text{Eq. 2.2}$$

where α is the probability of the logic gate output to change from 0 to1 and hence its value ranges from 0 to 1 and is called the switching activity. C_L is the load capacitance, and f_s is the switching frequency (i.e. clock frequency). The loading capacitance C_L mainly consists on the interconnecting parasitic capacitance, the gate capacitance of the subsequent stages, and the diffusion parasitic capacitance to ground of the drain of switching devices.

Due to increased demands on the system performance, the clock frequency increases. Power dissipation can be reduced by reducing the switching activity (Menon et al. 2004) and the output load capacitance. The former can be reduced via proper circuit and system designs, and the latter can be reduced by an advanced CMOS technology (input capacitance decreases with technology downscaling) or by reducing device dimensions. But the reduction in power dissipation is most effective when VDD is lowered in digital circuits (see section 2.4).

2.1.2 Leakage Current Power

There are two types of leakage currents: reverse-bias diode leakage on the transistor drains, and sub-threshold leakage through the channel of a turned-off device. The magnitude of these currents is set predominantly by the processing technology (leakage current increases with technology downscaling, Lin et al. 2002); however, there are some issues that designer can take into account in order to minimize their contribution(Hanchate and Ranganathan 2004), (Rahman and Chakrabarti 2004).

The power consumption due to leakage current can be calculated as follows:

$$P_{leakage} = \left(I_{diode} + I_{subthreshold}\right) VDD \quad \text{Eq. 2.3}$$

The diode leakage occurs when a transistor is turned off, and another active transistor charges up/down the drain with respect to theformer's bulk potential. In the case of the inverter with a high input voltage, the output voltage will be low because theNMOS transistor is on. The PMOS transistor will be turned off, but its drain-to-bulk voltage will be equal to the supply voltage, VDD. The resulting diode leakage current will be approximately $I_{diode} = A \cdot J$, where A is the area of the drain diffusion, and J is the leakage current density, set by the technology.

Since the diode reaches maximum reverse bias current for relatively small reverse bias potential, the leakage current is roughly independent of supply voltage. It is proportional to the diffusion area and perimeter; however, so it is desired to minimize the diffusion area and perimeter in the layout. The leakage current density is temperature sensitive, as well, so J can increase dramatically at higher temperatures.

On the other hand, subthreshold leakage occurs under similar conditions as the diode leakage. The magnitude of the subthreshold current is both a function of technology node, device sizing, and VDD. The process parameter that predominantly affects the current value is the threshold voltage of active devices (V_t). Reducing V_t exponentially increases the subthreshold current. For every transistor with diode leakage, the same bias conditions are present for subthreshold leakage, such that the total power dissipation of the two is roughly the same magnitude in both cases. The subthreshold current is also proportional to the transistor device size (W/L), and an exponential function of the supply voltage. Thus, the current can be minimized by reducing the transistor sizes, and by reducing the supply voltage.

2.1.3 Short-Circuit Current Power

During switching in CMOS circuits, both NMOS and PMOS transistors may be simultaneously active for a short period of time (Chatterjee et al. 1996), and therefore an instantaneous short-circuit current (I_{short}) flows from the power supply directly to ground. The power consumption due to the short circuit current is given by

$$P_{short} = I_{short} \cdot VDD \qquad \text{Eq. 2.4}$$

The peak magnitude of I_{short} current is dependent on device size. The average current, however, is roughly independent of device size for a fixed load capacitance. While the peak magnitude of the current increases, the rise/fall time decreases so that the average current is the same. If all devices are sized up so that the load capacitance scale up proportionally, then the rise/fall time remains constant and the average current (and power) scales up linearly with device size.

This term can be neglected if the signals have short rise and fall times as compared to duration of the signal.

2.1.4 Static Biasing Power

Commonly, static power consumption in CMOS circuits is only related to the leakage current (see 2.1.2). However, in analog circuits, static biasing current is the main contributor for power dissipation, as devices need to be biased permanently in the proper region (i.e. active region) of operation (see chapter 5). The static biasing power in analog circuits can be expressed as follows:

$$P_{bias} = I_{bias} \cdot VDD \qquad \text{Eq. 2.5}$$

where I_{bias} is the bias current. As said before, biasing power is dominant in analog circuits. That is the reason why the other means of power consumption described in 2.1.1, 2.1.2 and 2.1.3 can be neglected, and total power consumption of analog (and hence RF) circuits can be approximated by Eq. 2.5.

2.2 Limits in Power Dissipation

In this sections the limits of power consumption in CMOS circuits are described. Starting from a theoretical point of view, the fundamental limits are firstly pointed (see subsection 2.2.1), and secondly some practical limits are explained in subsection 2.2.2.

2.2.1 *Fundamental Limits*

Power is consumed in analog signal processing circuits to maintain the signal energy above the fundamental thermal noise in order to achieve the required signal-to-noise ratio (SNR). A representative figure of merit of different signal processing systems is the power consumed to realize a single pole (Enz and Vittoz 1997). The minimum power necessary to realize a single pole can be derived by considering a basic integrator assuming an ideal 100% current efficient transconductor (i.e. all the current pulled from VDD is used to charge the integrating capacitor, see Fig. 2.1).

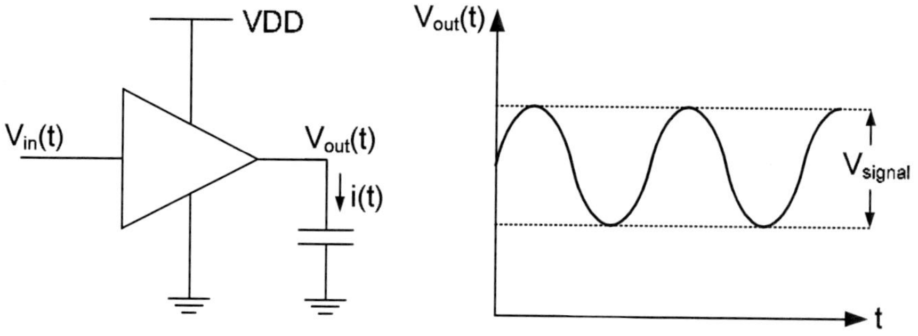

Fig. 2.1 100% current efficient transconductor for single pole realization

The power consumed from the supply voltage which is necessary to create a sinusoidal voltage V(t) across capacitor C having a peak-to-peak amplitude V_{signal} and a frequency f can be expressed as:

$$P = V_{DD} \cdot f \cdot CV_{signal} = f \cdot CV_{signal}^{2} \frac{V_{DD}}{V_{signal}} \qquad \text{Eq. 2.6}$$

whereas the signal-to-noise ratio is given by:

$$SNR = \frac{V_{signal}^{2}/8}{kT/C} \qquad \text{Eq. 2.7}$$

Combining Eq. 2.6 and Eq. 2.7 yields:

$$P = 8kT \cdot f \cdot SNR \frac{V_{DD}}{V_{signal}} \qquad \text{Eq. 2.8}$$

According to Eq. 2.8, the minimum power consumption of analog circuits at a given temperature is basically set by the required SNR and the operation frequency (or the required bandwidth). Since this minimum power consumption is also proportional to the ratio between $VDDV_{signal}$, power-efficient analog circuits should be designed to maximize the voltage swing. Therefore, the minimum power for circuits that can handle rail-to-rail signal voltages (V_{signal}=VDD) reduces to:

$$P_{\text{min,analog}} = 8kT \cdot f \cdot SNR \qquad \text{Eq. 2.9}$$

This absolute limit is very steep, since it requires a factor 10 of power increase for every 10 dB of improvement in SNR. It applies to each pole of any linear analog filter and is reached in the case of a simple passive RC filter, whereas the best existing active filters are still at least two orders of magnitude above.

On the other hand, the minimum power required for a voltage amplifier of gain $A_v$could be derived considering a single stage common-source (or common-emitter) small-signal amplifier (Enz and Vittoz 1997). In this study, it is concluded thatthe minimum power consumption for an amplifier is nA_v times larger than the limit given by Eq. 2.9.

In general, the minimum power for an analog system can be compared to that of a digital system, in which each elementary operation requires a certain number of binary gate transition cycles (m), each of which dissipates an amount of energy E_m. The minimum power is then simply given by:

$$P_{\text{min,digital}} = mBE_m \qquad \text{Eq. 2.10}$$

where B is the signal bandwidth. The number of transitions is only proportional to some power (a) of the number of bits N (N^a), and therefore power consumption is only weakly dependent on SNR (ref). Comparison with analog circuits might be obtained by estimating the number of gate transitions that are required to compute each period of the signal. Immunity to thermal noise imposes an absolute minimum energy per transitionestimated to 8kT, which provides the absolute minimum power limit However, in practice E_m is forced to a much higher value by the need to recharge the equivalent capacitance C of each gate to VDD. Therefore, the minimum power for digital is much higher than the absolute limitat room temperature.

As a conclusion, it can be stated that analog systems may consume much less power than their digital counterpart, provided that a small SNR is acceptable. But for systems requiring large SNRs (i.e. RF systems), analog design becomes very power inefficient.

2.2.2 *Practical Limits*

The limits discussed in the previous subsection are fundamental, as they do not dependeither on the technology or on the choice of supply voltage. However, a number of obstacles or technological limitations arise on the way to approach these limits in practical circuits:

- Capacitors increase the power necessary to achieve a given bandwidth. They are only acceptable if their presence reduces the noise power by the same amount (by reducing the noise bandwidth). Therefore, special care must be taken with parasitic capacitors, as they very often increase power consumption. This is a critical issue in RF circuits, where parasitic capacitances are more significant for higher frequencies.
- The power spent in bias circuitry is wasted and should be minimized. However, inadequate bias schemes may increase the noise and therefore require a proportional increase in power.
- The presence of additional sources of noise implies an increase in power consumption. These include 1/f noise in the devices, and noise coming from the power supply or generated on chip by other blocks of the circuit. For that reason, additional filtering and shielding is necessary to reduce power consumption in analog circuits.
- When capacitive loads are imposed (for example by parasitic capacitors), the current necessary to obtain a given bandwidth is inversely proportional to the transconductance-to-current ratio (g_m/I)of the active device. The small value of (g_m/I) inherent to MOS transistors operated in strong inversion may therefore cause an increase in power consumption. For that reason, subthreshold region operation (i.e. weak inversion) is preferred in low voltage applications (see subsection 2.4.2 and chapter 5).
- The need for precision usually implies the use of larger dimensions for active and passive components, with a resulting increase in parasitic capacitors and therefore power consumption.
- All switched capacitors must be clocked at a frequency higher than twice the signal frequency. The power consumed by the clock itself may be dominant in some applications.

2.3 VDD Downscaling

As mentioned in section 2.1, the total power consumption of a circuit can be approximated as the sum of both the dynamic and static sources of energy consumption.As pointed also in this section, both sources of power consumption are directly related to the supply voltage. As a consequence, the obvious way to reduce the total power consumption of any system would be to operate the circuits at lower supply voltages (Forestier and Stan 2000).

On the other hand, supply voltage reduction guarantee the reliability of devices as the lower electrical fields inside oxide layers of a MOSFET produce less risk to the thinner oxides, which result from device scaling. Thus, one of the solutions of all the problems lies in the adoption of low voltage techniques in analog circuit designs so that these MOSFETs can operate at low voltage levels.

As it has ben described in section 2.2, digital circuits benefit from supply voltage downscaling. However, when reducing the supply voltage, a number of limitations in the design and in the performance of analog circuits arise; for this reason, the next sections describe what the consequences and the impacts are with regard to the following issues:

- Threshold voltage
- Sub-threshold region
- MOS transistor speed
- Analog switches
- Transistor stacking
- Dynamic range
- Power consumption

2.3.1 Threshold Voltage

When supply voltage scales down, two important constraints are faced in analog CMOS design: the device noise level (Liu et al. 2006) and the threshold voltage(Sun and Tsui 1995), (Gonzalez et al. 1997). Reduction in threshold voltage is dependent on the device technology (see chapter 4). Higher threshold voltage values give better noise immunity, and lower values reduce the noise margin, leading to poor SNR. Hence, for modern CMOS technologies, reduction in the threshold voltage is limited to the noise floor level. Below this level, further threshold reduction introduces more noise in the circuit, leading to deal with very complex circuit techniques.

In digital and mixed-signal circuits, the inverter is a basic circuit, which implementation dictates the minimum supply voltage (VDD_{min}) required for a proper operation:

$$VDD_{\min} = V_{t,N} + |V_{t,P}| \qquad \text{Eq. 2.11}$$

where $V_{t,N}$ and $V_{t,P}$ are the threshold voltages for an NMOS and PMOS device respectively. As the gate of both transistors are tied together, if the supply voltage is lower that this limit, then a dead zone occurs in the middle of the input range.

Even when such configurations are avoided, in all analog (and hence RF) circuits the threshold voltage seriously limits the available signal swing. In fact, first of all, the power supply must have a minimum value so that a MOS device can be turned on; assuming strong inversion (see chapter 5), this condition can be expressed as

$$VDD - VSS \geq V_{GS} = V_{DS,sat} + |V_t| \qquad \text{Eq. 2.12}$$

In addition, if the transistor is gate-driven, the voltage swing of the input signal (V_{signal}) must be added to the previous power-on condition:

$$VDD - VSS \geq V_{GS} = V_{DS,sat} + |V_t| + V_{signal} \qquad \text{Eq. 2.13}$$

Finally, the power-on condition may be further restricted: let us consider one of the most basic configurations in analog circuits: the source-follower; in addition to the limit set by Eq. 2.13, the minimum supply limit requires headroom for at least one more drain-source saturation voltage. Assuming a 1.2 V power supply, a threshold voltage equal to 0.7 V (which is the typical value for a 3.3 V process) and a saturation voltage ($V_{DS,sat}$) of approximately 200 mV, the allowed signal swing is limited to at most 100 mV, under the assumption that the input signal is limited by the supply voltage. Hence, the threshold voltage is a strong limitation for the signal swing and, unfortunately, does not scale down at the same rate of the MOS channel length. However, further threshold voltage variation techniques might be applied in order to reduce the power consumption in CMOS circuits (see chapter 4).

2.3.2 *Sub-threshold Region*

When only small biasing currents are available (a desirable circumstance for low voltage operation), operating the MOS in saturation region could result in a smaller transconductance. The sub-threshold region (i.e. weak inversion) polarizes the gate of the transistor so that its gate-source voltage is below its threshold voltage (see chapter 5). In sub-threshold region, MOSFETs have lower saturation voltages ($\approx$ 100mV). This gives larger voltage swings at low-supply voltage even in cascaded MOSFET structures.

Another advantage of sub-threshold operation is a reduced input-referred noise contribution with respect to the saturation-region mode of operation, due to the larger transconductance. However, the relative output noise current is maximized, what prevents the use of sub-threshold MOS devices for biasing circuitry. Therefore, this mode of operation is desirable in analog (and RF) signal processing pahts.

On the other hand, several unwanted issues arise by using this technique:

- The lack of accuracy in setting the transistor current: the poor transistor matching limits the use of the sub-threshold region whenever current precision is required (as in current mirrors)
- Increased leakage currents (increasing the power consumption)
- Larger transistor sizes are required, increasing device parasitics (compared to the saturated MOS)
- Poorer frequency response of devices

- The drain and source substrate currents associated with the reverse biased moat-substrate junction are not necessarily negligible compared to sub-threshold drain current
- Finally, the linearity is quite poor for $V_{DS}<3V_{ther}$, where V_{ther} is the thermal voltage, given by (Lee 1998):

$$V_{therm} = \frac{KT}{q} \qquad \text{Eq. 2.14}$$

where K is the Boltzman constant, T is the absolute Temperature, and q is the electronic charge.

2.3.3 MOS Transistor Speed and Bandwidth

As pointed in section 2.2, reducing the supply voltage of analog circuits while preserving the same bandwidth and SNR has no fundamental effect on their minimum power consumption. However, this absolute limit was obtained by neglecting the possible limitation of bandwidth due to the limited transconductance of the active device. The maximum value of the bandwidth (B) is proportional to g_m/C, where C is the equivalent capacitance C of each gate to VDD.By replacing the capacitor value by g_m/Bin Eq. 2.7 and expressing the product of the SNR times the bandwidth yields:

$$SNR \cdot B = \frac{V_{signal}^2 g_m}{8kT} \qquad \text{Eq. 2.15}$$

In most cases, downscaling the supply voltage by a factor K requires a proportional reduction of the signal swing (V_{signal}).Maintaining the bandwidth and the SNR is therefore only possible if the transconductance is increased by a factor K^2. If the active device is a bipolar transistor (or a MOS transistor biased in weak inversion), its transconductance can only be increased by increasing the bias current (I) by the same factor (K^2); power consumption (VDD·I) is therefore increased by K. On the other hand, the situation is different if the active device is a MOS transistor biased in strong inversion. In this case, its transconductance can be shown to be proportional to I/V_p, where V_p is the pinch-off voltage of the device. Since V_p has to be reduced proportionally with VDD, increasing g_m by K^2 only requires an increase of current I by a factor K, and hence the power remains unchanged.

However, the maximum frequency of operation (and therefore the bandwidth) may be affected by the value of the supply voltage. For a MOS transistor in strong inversion, the frequencyfor which the current gain falls to unity (f_{max}) is approximately given by:

$$f_{\max} \cong \frac{\mu \cdot V_p}{L^2} \qquad \text{Eq. 2.16}$$

Therefore, if the technology node is fixed (channel length L constant) a reduction of *VDD* and V_p by a factor K causes a proportional reduction of f_{max}. Therefore, it seems that there is no fundamental reason to reduce the supply voltage of an analog circuit in a given process. On the other hand, a reduction of *VDD*is unavoidable to maintain the electric fields constant when scaling down a process. Both V_p and L are then scaled by the same factor (K), and the maximum frequency f_{max} is increased by K.

2.3.4 Analog Switches

The use of analog switches in low voltage design faces several challenges. A first obvious problem is the increased switch resistivity due to the reduced turn-on voltage available. To compensate this effect, transistors must be designed with larger dimensions, what leads to a larger clock feed-through and increased static power dissipation.

A non-conducting region centered around VDD/2 is exhibited by complementary switch when the voltage supply is below a critical voltage $V_{critical}$ given by (Enz and Vittoz 1997):

$$V_{critical} = \frac{2V_t}{2-n} \qquad \text{Eq. 2.17}$$

where n is the slope factor of the gate voltage versus pinch-off voltage. The critical voltage value $V_{critical}$ has been derived under the assumption that both NMOS and PMOS parameters of the complementary switch are the same.

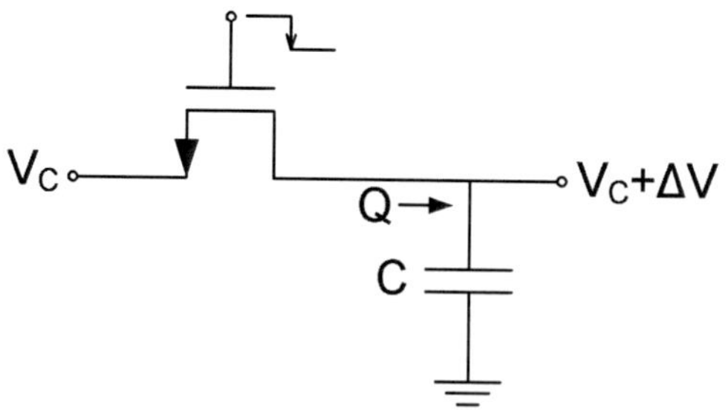

Fig. 2.2 Charge injection in analog switches

A well-known issue related to analog switches is the charge injection problem. When the switch is turned off, the charge in the MOS channel flows out from the channel region to the drain and source junctions. As illustrated in Fig. 2.2, the fraction of charge Q released into capacitor C causes an absolute voltage error equal to:

$$\Delta V = \frac{\Delta Q}{C} \qquad \text{Eq. 2.18}$$

The relative voltage error across the capacitor is given by:

$$\frac{\Delta V}{V} = \frac{\Delta Q}{VDD \cdot C} \qquad \text{Eq. 2.19}$$

where VDD·C is the capacitor maximum available charge. Eq. 2.19 indicates that the relative voltage error grows proportionally to the reduction of the supply voltage. This issue is critical in ultra-low-power systems, where power gating (see chapter 5) is used to power-down circuits in order to save power consumption.

2.3.5 Transistor Stacking

When the supply voltage is reduced, the allowed voltage swings of the circuit nodes become narrower, complicating the implementation of standard configurations such as stacked transistors.

For deep submicron CMOS technologies, the intrinsic transistor gain ($g_m r_{ds}$) is usually lower than 20 dB (Yaoet al. 2003). The most popular gain boosting configuration in analog circuits is the cascoding configuration (see Fig. 2.3) that, in low voltage design, is not readily available due to output swing limitations, as lower voltage headroom (V_{signal}) is left due to more than one stacked V_{DS} voltages ($V_{sat,1}$, $V_{sat,2}$).

On the other hand, cascoding could be replaced by cascading, i.e. folding (ref); however, cascading structures augment the power consumption and demand frequency compensation, since the gain boosting is achieved through several amplifying stages.

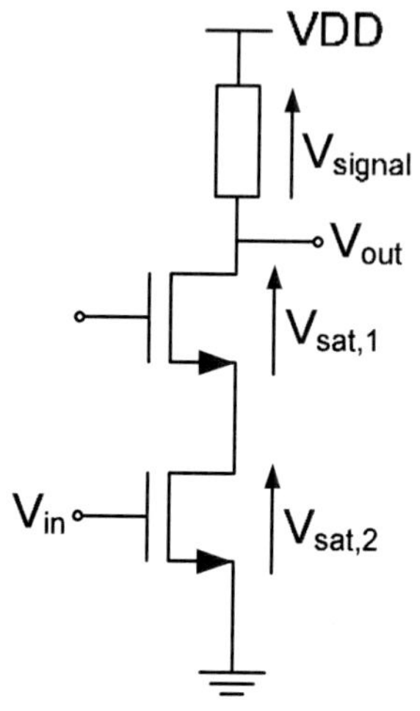

Fig. 2.3 Stacked transistors standard configuration

2.3.6 Dynamic Range

In this context, the dynamic ratio (DR) is defined as the ratio of the largest to the smallest possible signal amplitudes (the ratio of squared amplitudes is also a common definition). If we assume that DR is limited by noise on the lower side and by VDD on the upper side, the following expression is derived:

$$DR = \frac{V_{signal}/2}{V_{n,rms}} = \frac{\eta_v \cdot VDD}{2V_{n,rms}} \qquad \text{Eq. 2.20}$$

where V_{signal} is the peak-to-peak voltage, $V_{n,rms}$ is the rms value of the noise voltage, and η_v is the voltage efficiency defined by V_{signal}/VDD. If it is assumed that the lower side is limited by matching, then DR can be rewritten as

$$DR = \frac{V_{signal}/2}{\kappa \cdot \sigma(V_{GS})} = \frac{\eta_v \cdot VDD}{2\kappa \cdot \sigma(V_{GS})} \qquad \text{Eq. 2.21}$$

where κ is a yield parameter and $\sigma(V_{GS})$ is the standard deviation of critical offset voltage in the circuit. It can be concluded that, in any case, the supply voltage limits dynamic range: as VDD scales down, so does DR.

On the other hand signal-to-noise ratio (SNR) is defined as the ratio of the signal power to the noise power. For a sinusoidal signal, the maximum SNR is given by:

$$SNR = \frac{V_{signal}^2}{8V_{n,rms}^2} = \frac{(\eta_v \cdot VDD)^2}{8V_{n,rms}^2} \quad \text{Eq. 2.22}$$

what also indicates the dependence of SNR about the supply voltage: if VDD scales down, so does SNR. Combining eq. X and eq. Y the DR is directly related to the SNR:

$$DR = \sqrt{2 \cdot SNR} \quad \text{Eq. 2.23}$$

2.3.7 Power Consumption

It can be concluded that supply voltage downscaling does not significantly reduce the power consumption in analog circuits and introduces several design constraints, limiting the use of standard circuit configurations. The limits placed by the voltage supply downscaling are fundamental and can only be approached by proper design choices, usually at the expense of increased circuit complexity and power consumption. In particular, providing high gain and high output swing in analog amplifiers becomes very challenging in low-voltage applications; it appears clear that trade-offs between performance and power dissipation must be accepted as a natural consequence.

References

(Chatterjee et al. 1996) Chatterjee, A., et al.: An investigation of the impact of technology scaling on power wasted as short-circuit current in low voltage static CMOS circuits. In: International Symposium on Low Power Electronics and Design, August 12-14, pp. 145–150 (1996)

(Enz and Vittoz 1997) Enz, C., Vittoz, E.A.: MOS Transistor Modeling for Low-Voltage and Low-Power Analog IC Design. Microelectronic Engineering 39, 59–76 (1997)

(Forestier and Stan 2000) Forestier, A., Stan, M.R.: Limits to voltage scaling from the low power perspective. In: Proceedings of the 13th Symposium on Integrated Circuits and Systems Design, pp. 365–370 (2000)

(Gonzalez et al. 1997) Gonzalez, R., et al.: Supply and threshold voltage scaling for low power CMOS. IEEE Journal of Solid-State Circuits 32(8), 1210–1216 (1997)

(Hanchate and Ranganathan 2004) Hanchate, N., Ranganathan, N.: LECTOR: a technique for leakager eduction in CMOS circuits. IEEE Transactions on Very Large Scale Integration (VLSI) Systems 12(2), 196–205 (2004)

(Kiat-Seng and Kaushik 2005) Yeo, K.-S., Roy, K.: Low-Voltage, Low-Power VLSI Subsystems. McGraw-Hill, New York (2005)

(Lee 1998) Lee, T.: The Design of CMOS Radio Frequency Integrated Circuits. Cambridge University Press, Cambridge (1998)

(Lin et al. 2002) Lin, Y.-S., et al.: Leakage scaling in deep submicron CMOS for SoC. IEEE Transactions on Electron Devices 49(6), 1034–1041 (2002)

(Liu et al. 2006) Liu, M., et al.: Scaling Limit of CMOS Supply Voltage from Noise Margin Considerations. In: 2006 International Conference on Simulation of Semiconductor Processes and Devices, September 6-8, pp. 287–289 (2006)

(Menon et al. 2004) Menon, R.V., et al.: Switching Activity Minimization in Combinational Logic Design. In: Proceedings of the International Conference on Embedded Systems and Applications, pp. 47–53 (June 2004)

(Rahman and Chakrabarti 2004) Rahman, H., Chakrabarti, C.: A leakage estimation and reduction technique for scaled CMOS logic circuits considering gate-leakage. In: Proceedings of the 2004 International Symposium on Circuits and Systems, ISCAS 2004, May 23-26, vol. 2, pp. II-297–II-300 (2004)

(Sun and Tsui 1995) Sun, S.-W., Tsui, P.G.Y.: Limitation of CMOS supply-voltage scaling by MOSFET threshold-voltage variation. IEEE Journal of Solid-State Circuits 30(8), 947–949 (1995)

(Yao et al. 2003) Yao, L., et al.: A 0.8-V, 8-μW, CMOS OTA with 50-dB gain and 1.2-MHz GBW in 18-pF load. In: Proceedings of the 29th European Solid-State Circuits Conference, ESSCIRC 2003, September 16-18, pp. 297–300 (2003)

3 Impact of Architecture Selection on RF Front-End Power Consumption

A wireless receiver is typically composed by two sections: an analog front-end and a baseband digital processor. The analog section receives the modulated RF signal and downconvertes it to an appropriate intermediate frequency (IF) or directly to baseband (BB). The downconverted signal can be demodulated either in the analog domain or in the digital domain, after being digitized by an analog-to-digital (ADC) converter. The main issues related to this process are the following:

- Frequency translation of the wanted channel filtering out the unwanted interferers
- Lower noise contribution as possible in order not to degrade the signal's SNR
- The signal's integrity must be guaranteed
- Amplification of the signal's power up to the optimum level for the Analogue to Digital Converter (ADC) maximum efficiency

The interference-free frequency downconversion process is carried out by a mixer complemented by precise out-of-band and in-band channel filters and image rejection filtering. On the other hand, the minimum noise contribution is critical in the first stages of the front end (Friis 1945); therefore a Low Noise Amplifier (LNA) is key to keep the overall noise as low as possible. Nevertheless, the subsequent blocks need also to fit to the overall specification. For maintaining the integrity of the signal, a robust performance against blockers and self-distortion is guaranteed by means of increasing the dynamic range and linearity of each building block of the receiver, especially those at the last stages of the receiving path (Razavi 1998). Finally, the amplifying requirement depends on the power level detected at the RF input (sensitivity) and on the ADC's dynamic range: The analog (ready to be digitized) signal needs to go beyond the sensitivity level of the converter, but not further up to its saturation level.

3.1 Front-End Challenges

The first difficulty in the front end design is the image rejection. This is common to any architecture as unwanted signals exist in any configuration. Aside from this, the front end suffers from other problems within the signal processing, including

U. Alvarado et al.: Low Power RF Circuit Design in CMOS Tech., LNEE 104, pp. 25–40.
springerlink.com

even order distortion, DC offset, flicker noise and quadrature mismatch. In addition, both the sensitivity and linearity of the receiver must be maximized, while maintaining the power consumption as low as possible. All these issues affect the design of the blocks in their implementation and accuracy. A short overview of each of them is illustrated in the following subsections.

3.1.1 Image Rejection

In downconversion front ends using LO frequency different to RF, image frequency (IM) is also downconverted to IF band. Fig. 3.1 shows this phenomenon where the signal is corrupted by the image.

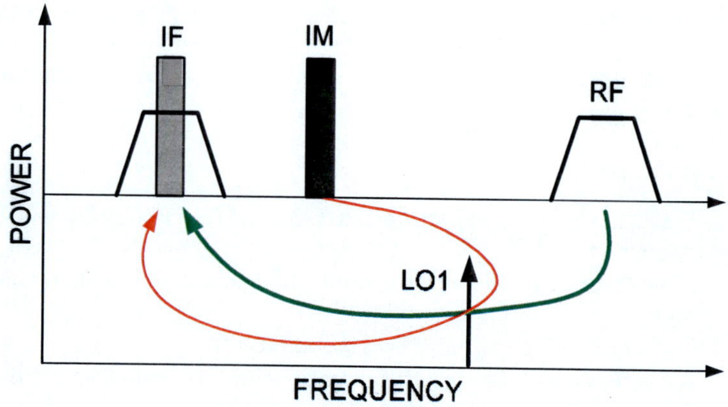

Fig. 3.1 Frequency Image problem

In order to avoid the undesirable effects of image frequency downconversion, IM shall be pre-filtered. Multiple discrete and integrated solutions have been found (Macedo and Copeland 1998), (Ray et al. 2000), but this represents a non affordable challenge when this image band falls into the vicinity of the RF band. In this case, polyphase filters are required (Crols and Steyaert 1998).

3.1.2 DC Offsets

When the signal is downconverted to baseband, the DC offset voltages that appear at the mixer output can corrupt the signal reducing its SNR (Fig. 3.2). Depending on its magnitude this voltage can saturate the following stages of the receiver. This is due to the fact that both the desired signal and the LO signals have exactly the same frequency. This issue is critical in direct conversion receivers (zero-IF).

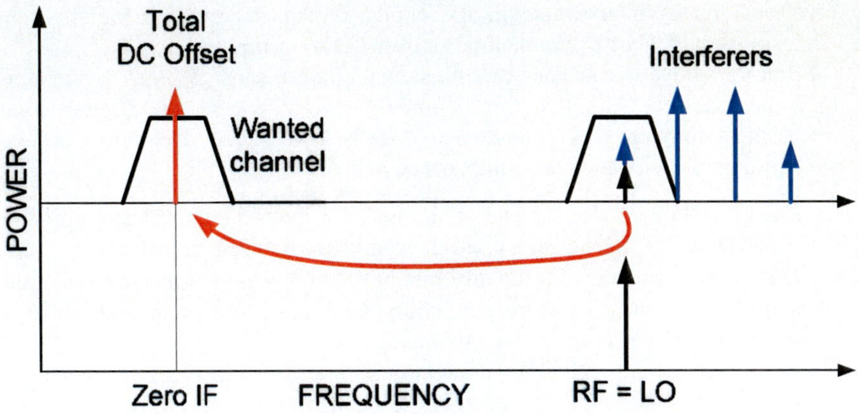

Fig. 3.2 DC offset in direct conversion receivers

In practical monolithic implementations the isolation between the LO port and the inputs of the mixer and the LNA is not infinite, and hence a finite amount of feedthrough exists from the LO port to the inputs of both the LNA and the mixer (see Fig. 3.3 (a). This effect is called *LO leakage* and it arises from capacitive and substrate coupling when the system is integrated in the same chip.

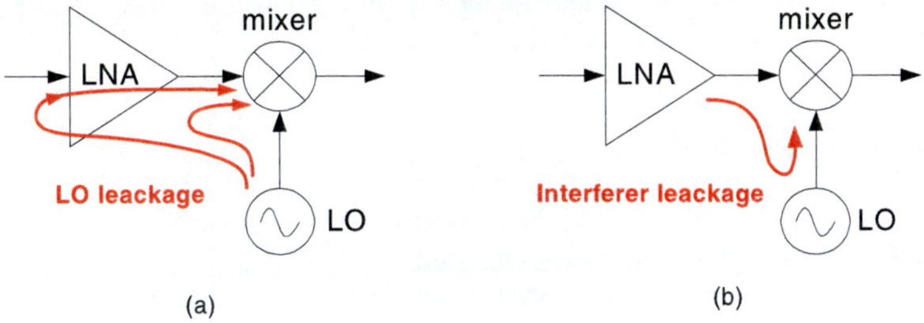

Fig. 3.3 Self mixing of (a) LO signal, (b) an interferer

The LO leakage is mixed with the LO signal, producing an unwanted DC component at the mixer output. This effect is called *self mixing*. A similar phenomenon occurs when a large interferer leaks from the LNA or the mixer input to the LO port and is mixed with itself (Fig. 3.3 (b). This DC offset is amplified by the remaining gain stages and hence early receiver overloading may occur, making the signal demodulation impossible.

The problem of DC offset becomes worse if self-mixing varies with time. This occurs when part of the LO signal leaks to the antenna and is radiated and then reflected from moving objects back to the receiver. Under these conditions, distinguishing between the time-varying offset and the actual signal may be difficult. Furthermore, leakage of the LO signal to the antenna and radiation therefrom creates interference in the band of other receivers using the same wireless standard. This phenomenon is called *LO re-radiation*.

From the above discussion, it is clear that some means of offset cancellation might be required. Although this is mainly carried out by the Digital Signal Processing subsystem (DSP), some steps can be taken when designing the front-end building blocks, which are summarized below.

- Layout techniques for increasing the isolation between component ports and the I/Q balance, taking care about the symmetry between branches.
- Differential structures for the different building blocks. In this case, the circuit symmetry is critical, therefore, component matching techniques such as common centroid must be used in the layout.
- Architecture of the mixers. An active mixer needs less LO signal power than its passive counterparts, therefore, the level of LO leakage is lower and hence the DC offset is reduced.
- Analog DC offset cancellation loops

Special care must be taken when designing the mixer. For a reduction of the DC offsets, a straightforward measure for its design is the choice of its architecture. Indeed, an active mixer needs less LO signal power than its passive counterparts. Therefore, the level of LO leakage is lower and hence the DC offset is reduced. DC offsets have been one of the reasons why zero-IF receivers have not been very popular for wireless applications in the past. Nevertheless, with a good DC offset cancellation performance carried out in the digital domain and a proper layout design this effect can be minimized.

3.1.3 I/Q Mismatch

Due to the modulation schemes used in DVB-H, the receiver must incorporate quadrature mixing. This requires shifting either the RF signal or the LO output by 90°. In both cases, the errors in the 90° shift and mismatches between the amplitudes of the I/Q signals corrupt the downconverted signal constellation (Fig. 3.4). All the stages in both I/Q paths contribute gain and phase errors.

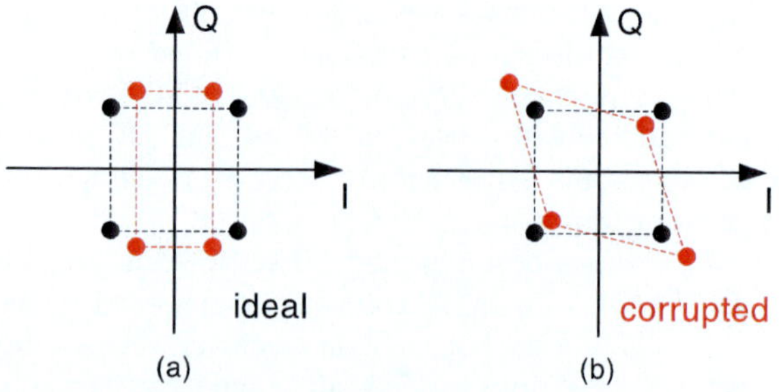

Fig. 3.4 Effect of I/Q mismatch on QPSK signal constellation. (a) Gain error, (b) Phase error

There are two different types of I/Q mismatch in a receiver with regard to gain and phase:

- Gain and phase constant mismatch, which can be easily corrected by means of the DVB digital demodulator.
- Gain and phase frequency dependent mismatch cannot be corrected as the previous mismatch. Therefore, this effect must be low enough so that it does not corrupt the required BER. In order to minimize this effect, special care must be taken in the design, especially regarding to layout techniques.

The problem of I/Q mismatching has been an obstacle in discrete implementations of direct conversion receivers, but it tends to improve with the monolithic integration. Indeed, different layout techniques such common-centroid contribute to improving the symmetry within components and hence the balances between I/Q paths. Furthermore, since mismatches vary negligibly with time, signal processing techniques may be utilized to correct the points in the constellation.

3.1.4 Even-Order Distortion

The effect of second order intermodulation can be quickly understood by examining a simple expression which relates the input and output signal of a building block through a high order transfer function. Firstly, assume that there is a non-linear transfer function relating the input and output baseband signals by

$$S_0(t) = a_1 S_i(t) + a_2 S_i^2(t) + a_3 S_i^3(t) + \ldots \qquad \text{Eq. 3.1}$$

where $S_0(t)$ is the output signal and $S_i(t)=S_i \cos(\omega_{in} t)$ represents the input signal at ω_{in} applied to an arbitrary building block. Using a simple trigonometric relationship for the second term the following expression is deduced

$$a_2 \left(S_i \cos(\omega_{in} t)\right)^2 = a_2 S_i^2 \left(\frac{1+\cos(2\omega_{in} t)}{2}\right) \qquad \text{Eq. 3.2}$$

It can be extracted from Eq. 3.2 so that when the input signal passes through a second order non-linearity a DC component is created. This is a serious problem when there is a weak desired signal that is frequency translated to baseband in the presence of a strong adjacent channel blocker. The large blocker now creates a DC component located in the center of the desired signal spectrum in the baseband.

The straightforward approach to suppress second order distortion is to implement differential LNA and mixers, but two considerations must be done. Firstly, the antenna and the band filter are usually single-ended. Thus, some means of converting the received signal to differential form is needed, for example transformers, which typically exhibit several dB of loss and contribute to raising the overall noise performance. Secondly, if the LNA is designed as a differential circuit, it requires higher power dissipation than a single-ended counterpart to achieve a comparable noise figure.

3.1.5 *Flicker (1/f) Noise*

The noise figure (NF) of the front-end determines the receiver sensitivity, that is, the ability of the receiver to demodulate the weakest of the signals. It is also known as white noise, since it is not dependent on the frequency. On the other hand, flicker noise in semiconductor devices, popularly known as *1/f noise*, is inversely proportional to the frequency (Fig. 3.5).

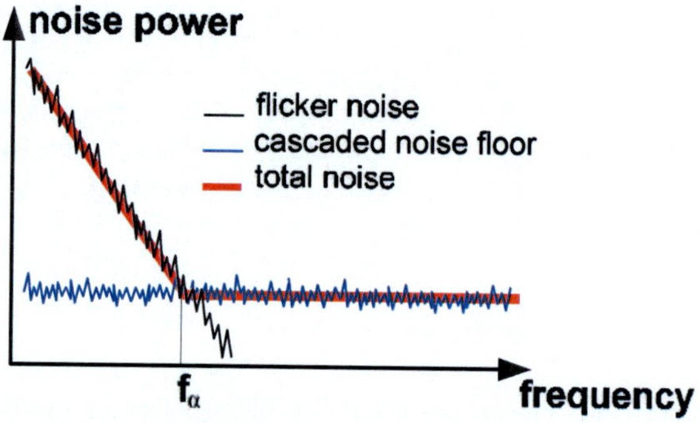

Fig. 3.5 Effect of flicker noise

Since the signals in a zero-IF front-end are directly downconverted to baseband, additional SNR degradation due to the flicker noise of the downconversion mixers could be significant. At input RF signals close to the receiver sensitivity level, the baseband I/Q signal at the mixers outputs is very low (usually in the range of tens of microvolts). The gain provided by the LNA and the direct-conversion mixer stages boosts both the signal and the noise floor.

The effective corner frequency f_α is a frequency where the flicker noise of the mixer equals the cascaded noise floor of the front-end, as shown in Fig. 3.5. Flicker noise property of a device is dependent on the technology process. f_α is typically in the range of 4~8 kHz for BiCMOS processes, while it can be up to 1 MHz for MOSFET devices (Lee 1998). Wideband systems like the DVB family signals have minimal energy at DC. Therefore, the SNR degradation due to flicker noise is lower than in narrowband systems.

The flicker noise effect can be minimized by the selection of the optimum semiconductor process with low f_α and providing adequate gain at RF to improve the relative SNR at the downconverter output. As the stages following the mixer operate at relatively low frequencies, they can incorporate very large devices to minimize the magnitude of flicker noise. This is due to the fact that 1/f noise is inversely proportional to device size (Lee 1998). Moreover, periodic DC offset cancellation in the analog front-end also suppresses low-frequency noise components through correlated double sampling.

3.1.6 Sensitivity and Noise Figure (NF)

The sensitivity of a receiver is the minimum detectable input signal (typically specified in dBm) such that there is a sufficient signal-to-noise-ratio at the output of the receiver ADC for a given application.

In order to estimate the sensitivity the following expression has been used (Razavi 1998):

$$P_{in,\min} = -174 dBm/Hz + 10\log BW + NF + C/N \quad \text{Eq. 3.3}$$

where BW is the channel effective bandwidth, NF is the total front-end noise figure and C/N is the required SNR for QEF demodulation. Note that the sum of the first three terms is the total integrated noise of the system and is called *noise floor*.

For receivers, the highest sensitivity as possible is desired, as it implies the detection of weaker signals. This is directly related with the front-end gain and noise figure: to reduce the NF improves the sensitivity, but on the other hand higher gain is needed to adequate the signal power to the ADC's dynamic range, which trades off the power consumption.

3.1.7 Linearity

Another major concern when designing the analog RF front-end parts of a receiver are the effects of the inherent nonlinearities of the active modules of the receiver chain. For any active device, the input/output voltage/current transfer characteristic is nonlinear, with saturation levels determined by the power supply voltage range. It is clear that for small signal levels, the behaviour is very linear, but with increasing signal amplitude, the nonlinearties have increasing impact. The linearity of a receiver directly affects the signal's integrity by degrading its SNR. Highly linear circuits are therefore desired within the entire receiving path, due to a double reason: circuit saturation and intermodulation.

On the one hand, if the input signal's power surpasses a circuit's saturation level (expressed by the 1-dB compression point) the circuit introduces distortion into the signal. As a consequence, non-desired frequency components appear at the output, degrading the signal's SNR. This effect is illustrated in Fig. 3.6, where a non-linear amplifier generated undesired frequency component at the output, reducing the signals SRN.

On the other hand, distortion is also produced even if the input signal's power is well below the circuit's saturation level. If other interfering signals at the vicinity of the desired signal are present at the input of a circuit, the resulting third order intermodulation products may corrupt the SNR. As illustrated in Fig. 3.7, two frequencies f_1 and f_2 will generate third order products at ($2f_1$-f_2) and ($2f_2$-f_1). Third order inter-modulation products can occur in receivers when two or more strong signals are present. These generate unwanted mixing signal products which

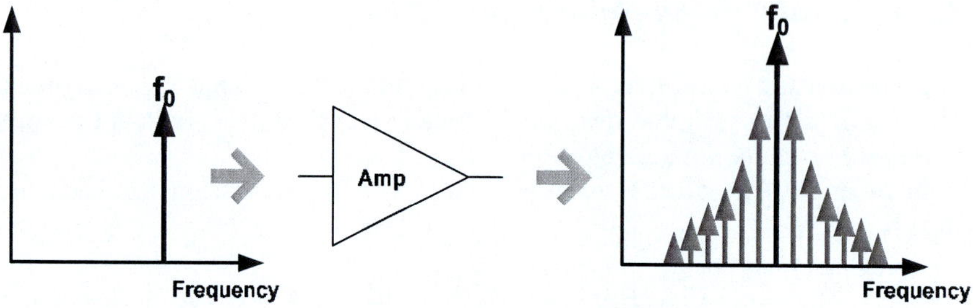

Fig. 3.6 Distortion created by a non-linear amplifier

may coincide with the operating frequency of the receiver, as depicted in Fig. 3.7, where $2f_1$-f_2 product falls over the desired signal, degrading its SNR. This, in turn, may cause the desired signal to be "masked" unless it is also fairly strong. In general there will always be a higher level at which a single unwanted signal will saturate the front-end of the receiver, thereby de-sensing it. This effect is referred to as "blocking."

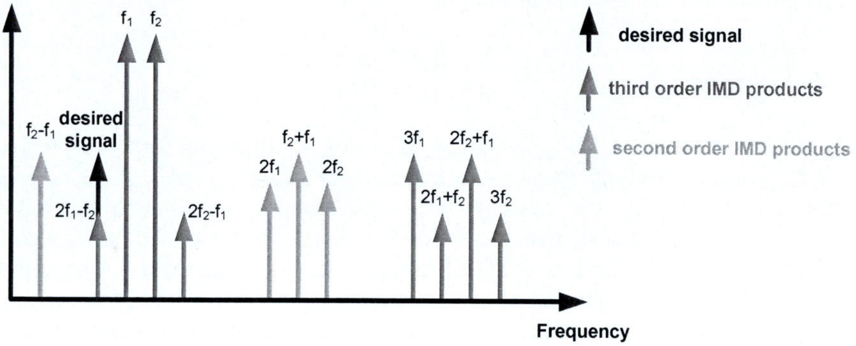

Fig. 3.7 Intermodulation (IMD) products

For these reasons, it is crucial to maximize the front-end's linearity, especially in the latest stages of the receiving chain, where the power of the signal is close to the overload levels. The following hints might be applied:

- In order to avoid out-of-band interferers, higher Q filtering is necessary at RF.
- In order to design more linear circuits, some kind of linearization loops might be implemented (Kim et al. 2004), (Cha et al. 2009).
- To avoid overloading, it is very interesting to implement automatic gain control loops, in order to adapt the signal power to the dynamic range of the subsequent stage.

3.2 Superheterodyne Architecture

Traditional receivers use a super heterodyne architecture known as "single conversion" (Fisher et al. 1952), (Samavati et al. 2000), for the analog front-end implementation. The input RF signal is downconverted to a low IF by means of a mixer, while unwanted signals are filtered out at both RF (out-of-band interferers) and IF (image frequency) to achieve the desired performance. A block diagram of this architecture is shown in Fig. 3.8.

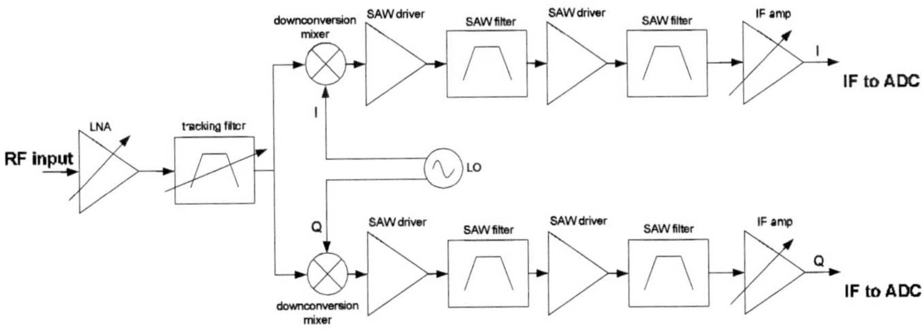

Fig. 3.8 Low-IF single conversion (superheterodyne) front-end simplified block diagram

The RF filter needs to be centered on the wanted channel, and therefore must be able to track the desired frequency when the user selects a given channel. This characteristic makes it difficult to integrate and it is usually implemented with discrete external components such as air coils and variable capacitors.

On the other hand, the IF filter is more selective than the band filter at RF. Two SAW filters are commonly used for this purpose in superheterodyne receivers to achieve the wanted rejection of interferers. These filters are expensive and introduce losses that degrade the overall noise performance of the system. In addition, they cannot be integrated and must be added as external components. As a result of its high number of external components such tuners are usually produced as a stand-alone component in a can, being generally bulky and too large for portable terminals.

3.3 Double Conversion Architecture

To overcome the size factor and to relax the selectivity performance at RF, the concept of heterodyning can be extended to multiple frequency translations. In this architecture, "double conversion", there are no tracking filters at RF before the mixer. A first mixer up converts the band of interest into a first fixed frequency (1st IF), at which a wide SAW filters performs a rough image rejection. A second frequency conversion downconverts the desired channel into a low IF (2nd IF), where the channel filtering is performed by means of two SAW filters, as shown on Fig. 3.9.

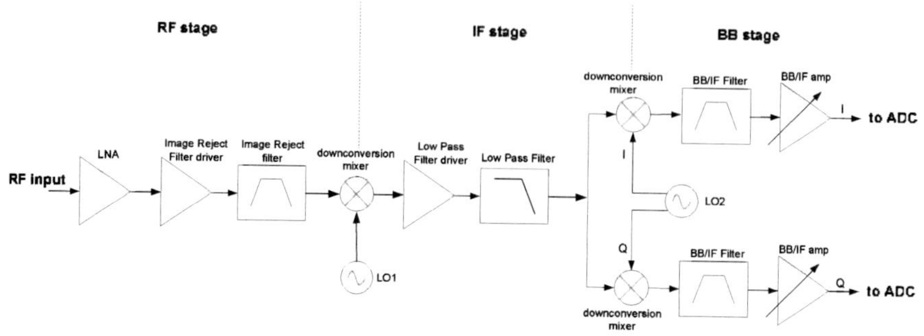

Fig. 3.9 Double conversion front end architecture

The RF signal is amplified by the LNA and usually prefiltered off chip by an image reject filter. This provides a clean first down conversion to an intermediate frequency IF = RF – LO1. Once in the IF band, a low pass filter eliminates the up conversion component generated (RF + LO1). Then, the signal is separated in its quadrature terms for the second mixing stage, by means of a second oscillator signal LO2. Finally, a channel filter selects the focused channel which is amplified before the Analogue to Digital Converter.

The election of the IF has three different possibilities, whose pros and cons are outlined in Table 3.1. This decision lays on a trade-off among the frequency of the synthesizer, the selectivity and feasibility of the IF band filters, and the coupling of the powers here implied.

Table 3.1 IF selection considerations for double conversion architectures

IF selected	ADVANTAGES	DISADVANTAGES
High IF (Antoine et al. 2005)	Image band easily filterable as it is far away from the RF band	Higher frequency for the IF filter
IF = RF/2 (Razavi 2001)	Only one Synthesizer Image band at 0 Hz	IF / LO coupling
Low IF (Lee 2002)	Better selectivity in the IF band	High LO1 frequency

With regard to power consumption considerations, it should be noted this architecture requires a high number of blocks. The image reject filter and low pass filter are usually external components after the first down conversion. Consequently, the signal needs to be driven out of the chip before them, which in turn requires higher currents. The need of two synthesizers also increases the die area and the power consumption and so is not recommended for low power applications.

3.4 Image-Rejection (Hartley, Weaver) Architecture

The image frequency problem appears as one of the most costly problems of the heterodyne front ends. The most extended specific solutions are based on the strongest filtering or on particular architectures with selective mixing, as outlined in the following subsections.

3.4.1 Hartley Architecture

The architecture presented in Fig. 3.10 (Hartley 1928) mixes the RF differential signal with the quadrature phases (cosine and sine) of the LO. The result is low-pass filtered and one branch per phase is shifted by 90°. At this point, before the adder the signal components have the same polarity whereas the image components have opposite polarity, and this is why the result of each addition is the downconversion of the RF input with no corruption from the image.

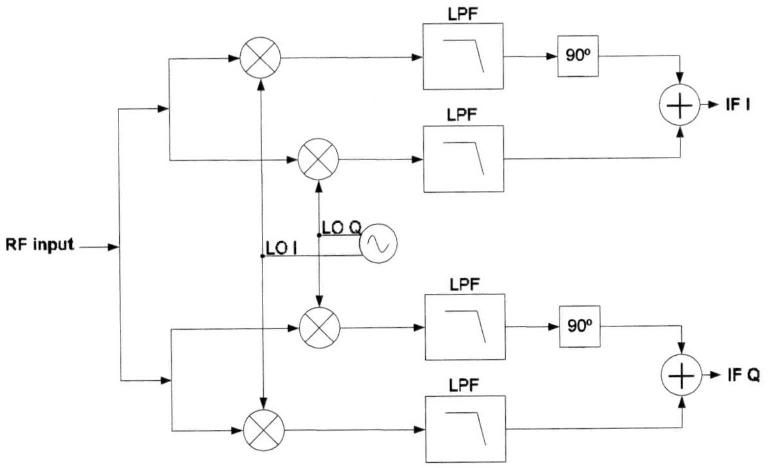

Fig. 3.10 Quadrature Hartley architecture block diagram

The principal drawback of this architecture is its sensitivity to mismatches. If the LO phases are not in exact quadrature or the gains and phase shifts of the upper and lower paths are not identical, the cancellation of the image is not complete. In this case, the down converted signal is corrupted. Besides, the necessity of a 90° phase shifter in only one path of each branch presumes the addition of mismatch to the operation described by (Hartley 1928).

Monolithic implementation of the Hartley architecture entails other issues as well. First, since the low-pass filters in Fig. 3.10 cannot easily suppress strong interferers in the neighbouring channels, the linearity of the adder is critical. This imposes additional noise-power trade-offs. Second, the loss and the noise of the phase shift stage are quite significant. And finally, the high power consumption due to the number of circuits makes this architecture inadvisable for the low power aim of this study.

3.4.2 *Weaver Architecture*

This architecture (Weaver 1956), shown in Fig. 3.11, implements additional mixers in order to produce the same polarities for the desired signal paths and opposite polarities for the image's ones, as well as the Hartley architecture proposed. The 90° phase shifter is here avoided by the usage of a second quadrature mixing operation, which at least avoids the one-sided mismatches before the adder present in Hartley's architecture.

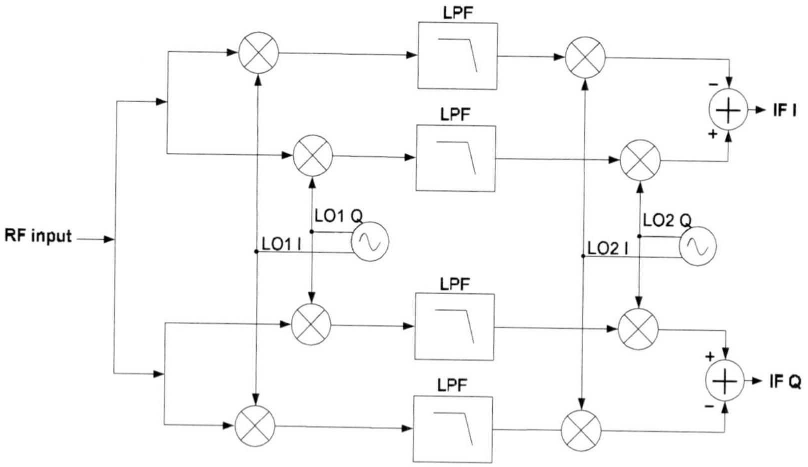

Fig. 3.11 Quadrature Weaver architecture block diagram

Nevertheless, this architecture does not perform a complete image rejection due to gain and phase mismatches. It suffers as well from the secondary image, if the second down conversion translates the spectrum to a nonzero frequency. Besides, harmonics of LO2 may downconvert unfiltered interferers from the intermediate stage to the output of this architecture. In any case, the number of circuits, and hence the even higher power consumption than the Hartley solution, makes this architecture inappropriate for its implementation in handheld devices.

3.5 Direct Conversion Receiver Architectures

Direct conversion architectures (Razavi 1997), (Namgoon 2001) benefit from higher integrability compared to the classical dual-IF structure. No external filters are needed for image rejection or for intermediate band selection. This avoids the external insertion loss and the extra power consumption from the drivers. Furthermore, the elimination of external blocks increases the robustness of this architecture, since the interaction with the parasitics of the package is minimized. The decreasing of the sensitivity to interferences and crosstalk is one of the main advantages of this compact proposal. The single down conversion stage is one of the strongest arguments for the consideration of this architecture for minimal power consumption.

Regarding the image problem, it does not disappear but it is translated to the baseband chain, where it can be solved without costly high Q high frequency filters. The image band is, in this case, the surrounding of the wanted channel, or the wanted channel itself as in the specific case of zero IF conversion. The image cannot be filtered before mixing but after its quadrature down conversion.

Two architectures are possible for single conversion. Here they are presented highlighting their characteristics in order to make an appropriate comparison.

3.5.1 Zero IF Architecture

A simplified zero-IF front end (Ismail and Abidi 2005) (Duvivier et al. 2003) is shown in Fig. 3.12. The LO frequency is equal to the input centre frequency of the wanted channel. Hence, the spectrum is translated to the baseband by means of a direct down conversion.

The simplicity of this architecture offers two important advantages over the others presented above. Firstly, as the image frequency is the proper signal the image problem disappears and therefore no image filter is required. As a result, the LNA does not need to drive a 50Ω load, making its implementation easier. Secondly, the IF SAW filter and subsequent downconversion stages are replaced with low-pass filters and baseband amplification stages, which are amenable to monolithic integration.

The simplicity of this architecture, due to the lesser amount of components (in comparison with traditional super-heterodyne architectures) causes two important consequences. One of the main causes is the reduction of the size and the cost of the receiver, because the number of external components is smaller. Also lower power consumption, makes this receiver suitable for hand-held battery-powered portable devices.

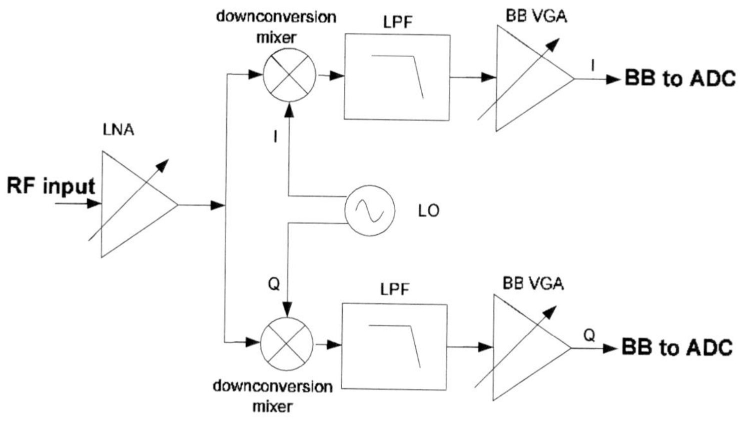

Fig. 4.7:

Fig. 3.12 Zero IF architecture block diagram

However, the direct conversion to zero frequency entails a number of issues that do not exist or are not as critical in other receiver architectures, such as DC offsets, I/Q mismatch, even-order distortion, flicker noise and LO re-radiation, as described in section 3.1.

3.6 Low IF Architecture

The non Zero IF alternative for the direct conversion architectures is the Low IF structure (Adiseno et al. 2002). The block composition is depicted in Fig. 3.13. It maintains the high integration benefits whereas it eludes the even order distortion, the DC offsets and the LO leakage problems. The desired channel is typically downloaded to the few MHz region. Distortion and self-mixing effects do not affect the signal, and consequently do not saturate the baseband stages. With a careful selection of the IF frequency location, this architecture avoids the 1/f flicker noise problems. The mixer design would here be relaxed in what concern noise and minimum LO power needed. This helps to lower the power consumption of this block.

Nevertheless, the direct conversion architectures share the problem of the I/Q mismatch since the quadrature data for the demodulation are needed. Low IF, furthermore, requires the I/Q components for the filtering. The polyphase filter after the down conversion is one of the main particularities of this configuration. The image band of this structure is only few MHz away from the desired band and consequently has not the possibility to be filtered before the signal treatment. This is done through operations with the down converted quadrature components of the signal, and performs an extremely sharp selectivity.

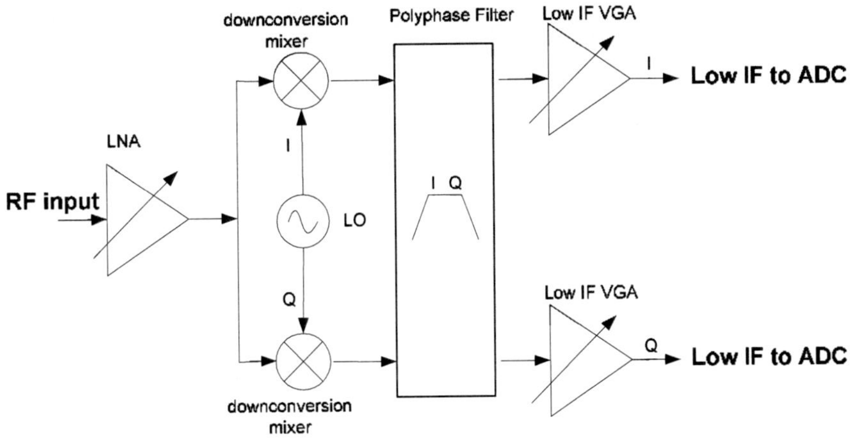

Fig. 3.13 Low IF architecture block diagram

The digital conception of the filter plays an important role in the multi standardization of the recent front ends. It would be easy to reconfigure it in case of variation of the unwanted frequency band, or change it as can do the wanted frequency bandwidth. Furthermore, this configuration has been presented as a useful tool in the MIMO receivers (Rafati and Razavi 2007). As described in this reference, the polyphase filter operation allows the reception of two simultaneous signals, as it leads them separately to the ADC, changing their sign. Another advantage of the digitalization of this element is the direct scalability in case of down scaling in the CMOS technology selected for its implementation.

References

(Adiseno et al. 2002) Adiseno, Ismail, M., Olsson, H.: A wide-band RF front-end for multiband multistandard high-linearity low-IF wireless receivers. IEEE Journal of Solid State Circuits 37(9), 1162 (2002)

(Antoine et al. 2005) Antoine, P., et al.: A direct-conversion receiver for DVB-H. IEEE Journal of Solid-State Circuits 40(12), 2536–2546 (2005)

(Cha et al. 2009) Cha, C.-Y., Lee, H.-B., Kenneth, K.O.: A TV-Band Harmonic Rejection Mixer Adopting a g_m Linearization Technique. IEEE Microwave and Wireless Components Letters 19(9), 563 (2009)

(Crols and Steyaert 1998) Crols, J., Steyaert, M.S.J.: Low-IF topologies for high-performance analog front ends of fully integrated receivers. IEEE Transactions on Circuits and Systems II: Analog and Digital Signal Processing 45(3), 269–282 (1998)

(Duvivier et al. 2003) Duvivier, E., Puccio, G., Cipriani, S., Carpineto, L., Cusinato, P., Bisanti, B., Galant, F., Chalet, F., Coppola, F., Cercelaru, S., Vallespin, N., Jiguet, J.-C., Sirna, G.: A fully integrated zero-IF transceiver for GSM-GPRS quad-band application. IEEE Journal of Solid State Circuits 38(12), 2249 (2003)

(Fisher et al. 1952) Fisher, D.H., Segrave, P.A., Watts, A.J.: The design of a superheterodyne receiver for television. Proceedings of the IEE - Part IIIA: Television 99(19), 609 (1952)

(Friis 1945) Friis, H.T.: Discussion on Noise Figures of Radio Receivers. Proceedings of the IRE 33(2), 125 (1945)

(Hartley 1928) Hartley, R.: Single-sideband modulator, U.S. Patent 1 666206 (April 1928)

(Ismail and Abidi 2005) Ismail, A., Abidi, A.A.: A 3.1- to 8.2-GHz zero-IF receiver and direct frequency synthesizer in 0.18-μm SiGe BiCMOS for mode-2 MB-OFDM UWB communication. IEEE Journal of Solid State Circuits 40(12), 2573 (2005)

(Kim et al. 2004) Kim, T.W., Kim, B., Lee, K.: Highly linear receiver front-end adopting MOSFET transconductance linearization by multiple gated transistors. IEEE Journal of Solid State Circuits 39(1), 223 (2004)

(Lee 1998) Lee, T.: The Design of CMOS Radio Frequency Integrated Circuits. Cambridge University Press, Cambridge (1998)

(Lee et al. 2002) Lee, T.H., Samavati, H., Rategh, H.R.: 5-GHz CMOS wireless LANs. IEEE Transactions on Microwave Theory and Techniques 50(1), 268 (2002)

(Macedo and Copeland 1998) Macedo, J.A., Copeland, M.A.: A 1.9-GHz Silicon Receiver with Monolithic Image Filtering. IEEE Journal of Solid-State Circuits 33(3), 378–386 (1998)

(Namgoong 2001) Namgoon, W., Meng, T.H.: Direct-conversion RF receiver design. IEEE Transactions on Communications 49(3), 518 (2001)

(Rafati and Razavi 2007) Rafati, H., Razavi, B.: A Receiver Architecture for Dual-Antenna Systems. IEEE Journal of Solid State Circuits 42(6), 1291 (2007)

(Ray et al. 2000) Ray, B., Hamel, J.S., Manku, T., Nisbet, J.J.: A Highly Selective Passive Band Reject Filter with Low-Q Lumped Elements in a Si Bipolar Process. In: IEEE BCTM (2000)

(Razavi 1997) Razavi, B.: Design considerations for direct-conversion receivers. IEEE Transactions on Circuits and Systems II: Analog and Digital Signal Processing 44(6), 428 (1997)

(Razavi 1998) Razavi, B.: RF Microelectronics. Prentice Hall Inc., Englewood Cliffs (1998)

(Razavi 2001) Razavi, B.: A 5.2-GHz CMOS receiver with 62-dB image rejection. IEEE Journal of Solid-State Circuits 36(5), 810 (2001)

(Samavati et al. 2000) Samaviti, H., Rategh, H.R., Lee, T.H.: A 5GHz CMOS Wireless LAN Receiver Front End. IEEE Journal of Solid-State Circuits 35(5), 765–772 (2000)

(Weaver 1956) Weaver, D.K.: A third method of generation and detection of single-sideband signals. In: Proc. IRE, vol. 44, pp. 1973–1705 (1956)

4 Technology Structural Alternatives in Standard CMOS Technologies for Low-Power Analog Design

As it is well known by electronic designers, each fabrication process has its own design rules and specific features. However, there are some alternatives usually provided by almost every standard foundry, presented and discussed in this chapter. Section 4.1 explains Variable Threshold CMOS devices and the use of multi-threshold transistors to reduce power consumption. The different body biasing alternatives are shown in Section 4.1. Gate length downscaling is a constant trend in modern CMOS processes, its impact being crucial in several key parameters of RF circuit performance. Therefore in Section 4.2 a discussion on the different cross-influences is presented. Finally, Chapter 4 closes with a discussion of the benefits of SOI processes.

4.1 Threshold Voltage (V_T)

The threshold voltage of CMOS transistors may have a great impact in the power consumption of a circuit. Switching devices (e.g. digital systems, frequency dividers, mixers) show a better performance in terms of switching efficiency for lower threshold values (Alvarado et al. 2010), reducing the dynamic power consumption. In addition, static leakage power consumption also depends on the threshold of transistors (Narendra et al. 2002). However, the performance of some other transistors might be degraded when reducing their threshold, especially if operating under low supply voltage conditions (Langen and Huijsing 1998). Therefore, multiple-threshold transistors in a single chip appear as the best way to optimize the power consumption of a circuit. This way, multiple-threshold devices (i.e. both high and low threshold transistors, MTCMOS) and variable-threshold devices (VTCMOS) are available in some CMOS technologies.

4.1.1 Multiple-Threshold Transistors

Multiple-threshold transistors (MTCMOS) were first proposed by (Mutoh et al. 1996) by inserting high threshold devices in series into low threshold circuitry. Shown in Fig. 4.1 the schematic of a general MTCMOS circuit, where sleep control (e.g. power-up / power-down) scheme is introduced for a more efficient power management. This way, in the active mode of operation the SLEEP signal is set to low, and sleep control high threshold devices are turned on. Since their on-state resistances are small, the virtual supply voltages (VDD and GND) act as

U. Alvarado et al.: Low Power RF Circuit Design in CMOS Tech., LNEE 104, pp. 41–60.
springerlink.com

real supply power rails. In the sleep mode, SLEEP is set to high, and both CMOS transistors in Fig. 4.1 are turned off. As their threshold voltage is high, the leakage current is low, and hence the power consumption is reduced.

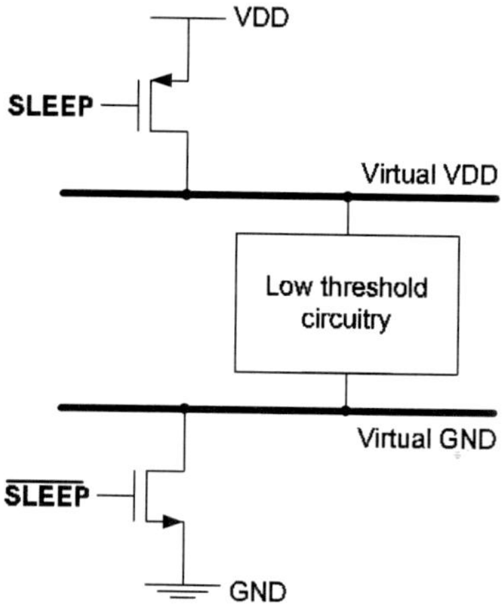

Fig. 4.1 Simplified schematic of a MTCMOS circuit

In fact, only one type of high threshold transistor (either PMOS or NMOS) is needed to perform the operation described above. Depicted in Fig. 4.2 are two different implementationsusing PMOS and NMOS devices respectively. An NMOS insertion scheme is preferable, as the NMOS on-resistance is lower than that of a PMOS device for the same size.

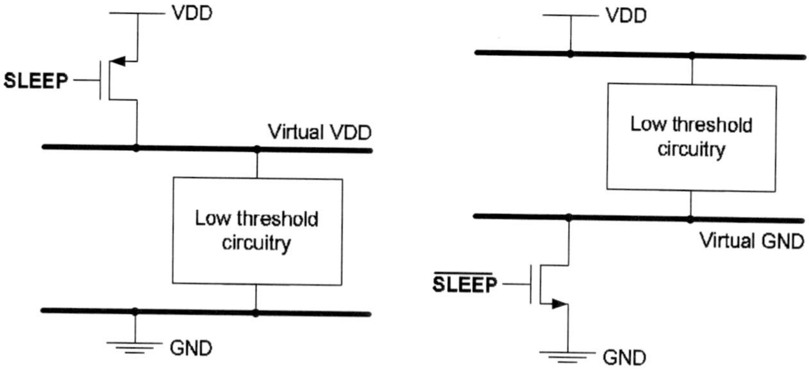

Fig. 4.2 Alternatives for a MTCMOS circuit

However, the implementation of high-threshold switches with low on-state resistance implies big sized transistors, what may increase the die area and delay. In this case, super cut-off transistors (SCCMOS) can be used with an inserted bias generator (Kawaguchi et al. 1998), to fully cut off the leakage current. This technique allows working with lower supply voltages and is commonly used in digital circuits.

RF circuits may benefit from MTCMOS techniques. For example, the switching stage in CMOS mixers is directly related with the flicker noise (Alvarado et al. 2010), what is a critical aspect in the design of direct conversion (zero-IF) receivers. Low threshold transistors can be used as switches to increase the switching efficiency, while higher threshold is preferred in the trasconductance stage. Other situation where MTCMOS techniques are desired is in ultra-low power wace-up receivers (Zhiyu and Kursun 2008), or in RFID systems, where ultra-low threshold devices are needed to boost the efficiency of the voltage rectification stage (Gosset et al. 2008).

In CMOS technologies, multiple-threshold devices can be obtained by means of the following techniques:

- Multiple channel doping
- Multiple oxide CMOS (MOXCMOS) circuits
- Multiple channel length
- Multiple body bias

4.1.1.1 Multiple Channel Doping

By changing the doping densities of transistors' channels multiple-threshold devices can be obtained, as shown in Fig. 4.3 (Kiat-Seng and Kaushik 2005), where the results are based on theoretical simulations. In order to have more than a single doping density, two additional masks are required with regard to traditional CMOS. However, this process entails a significant dispersion between different dies, and even within the same die, especially for different threshold values close to each other. For this reason, foundries usually provide only two (or three) different thresholds, which values are separated enough in order to cope with such process deviations.

4.1.1.2 Multiple Oxide CMOS (MOXCMOS) Circuits

The threshold voltage of CMOS transistors is also related to the gate-oxide thickness (t_{ox}). This way, by using large gate-oxide thicknesses higher threshold voltages are obtained. With larger oxide thickness the gate capacitance is reduced, and therefore both dynamic and subthreshold leakage power may be reduced, as well as the gate-oxide tunnelling leakage (Chang et al. 1998). On the contrary, lower threshold devices can be obtained by reducing the gate-oxide thickness, for a better efficiency and speed in switching devices. However, the process technology for multiple gate-oxide thicknesses is complicated.

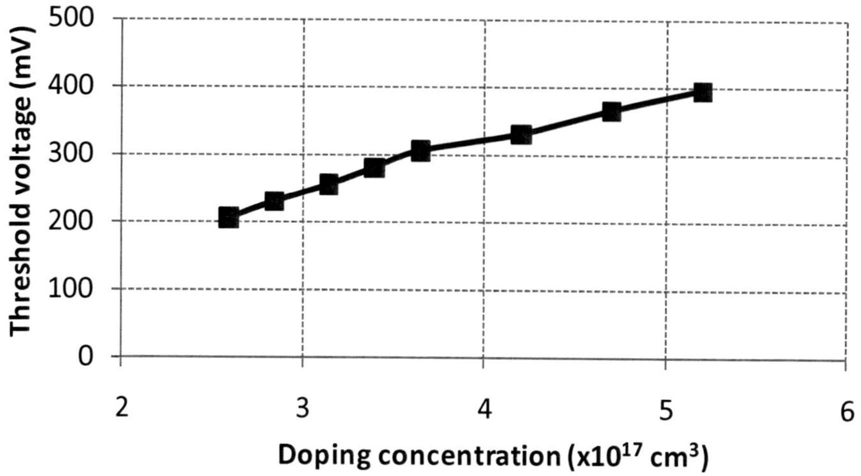

Fig. 4.3 Threshold voltage at different channel doping densities

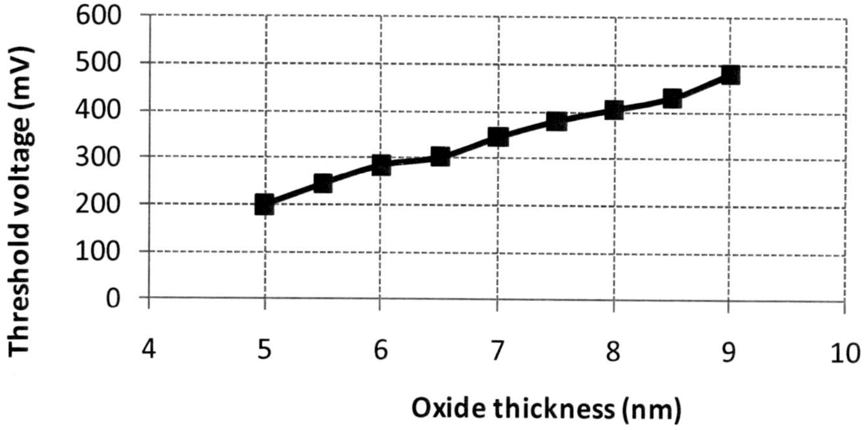

Fig. 4.4 Threshold voltage at different gate-oxide thicknesses

An important issue regarding to this technique is related to the aspect ratio (AR) of transistors, i.e. the ratio between the transistor's lateral and vertical dimensions, what measures short-channel immunity. To suppress the short-channel effect requires increasing the transistor's channel length in order the device to present a good aspect AR, given by the following equation (Kiat-Seng and Kaushik 2005)

$$AR = \frac{L}{\sqrt[3]{t_{ox} d \left(\varepsilon_{si} / \varepsilon_{ox}\right) X_j}} \quad \text{Eq. 4.1}$$

Where L is the channel length, t_{ox} is the oxide thickness, ε_{si} and ε_{ox} are the permitivities of silicon and oxide respectively, d is the depletion depth and X_j is the junction depth. It is a good strategy to change the gate-oxide thickness along with the channel length in order to keep a constant AR. Fig. 4.4 shows the variation of the threshold voltage vs. gate-oxide thickness for a constant AR.

4.1.1.3 Multiple Channel Length

For short channel devices, voltage threshold decreases as technology scales down and therefore with channel length downscaling (threshold roll-off). Depicted in Fig. 4.5 is the voltage threshold vs. channel length for different values of drain-source voltages.

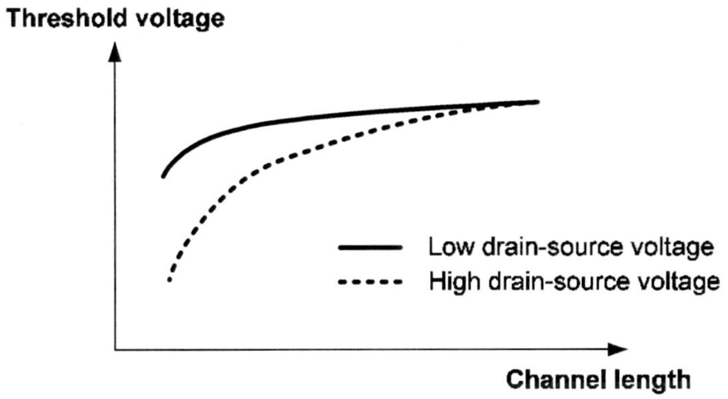

Fig. 4.5 Threshold voltage roll-off with change in channel length

Starting from a minimum value depending on the CMOS technology (e.g. 90 nm, 45 nm) the channel length of any transistor may be easily changed without any processing extra feature. On the other hand, when channel length is close to 100 nm, Halo doping profiles (Andricciola and Tuinhout 2009) are necessary to suppress the short-channel effect. For such doping profiles, the threshold roll-off might be too sharp and hence difficult to control near the minimum feature size for the used technology.

4.1.2 Variable-Threshold Transistors

Variable threshold CMOS (VTCMOS) is a body biasing design technique. The main objective of this technique is to reduce the leakage current of circuits, in order to reduce the power consumption in sleep or idle mode of operation. In order to achieve different threshold voltages, a self-substrate bias circuit is applied (see Fig. 4.6) to control the body bias of active devices:

- In the active mode, zero body bias is applied. In such case, the bulk of PMOS devices is connected to VDD while the bulk of NMOS devices is connected to VSS, reducing the threshold of transistors and therefore reducing the dynamic power consumption (e.g. in high speed switching circuits such as frequency dividers or VCOs. Depending on the circuit, a slightly forward substrate bias can be used to increase the speed while reducing the short-channel effect (Oowaki et al. 1998).
- In the standby mode, a deeper reverse body bias is applied in order to increase the threshold voltage of devices so that the cut off leakage current is reduced. This way, the power consumption of the circuit in the sleep mode of operation is reduced.

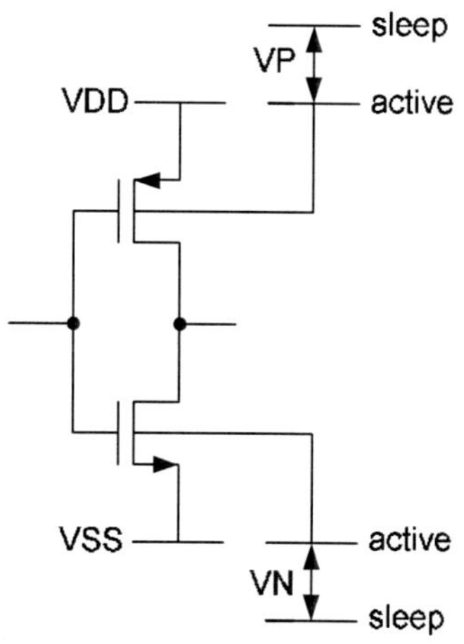

Fig. 4.6 Variable threshold CMOS (body biasing)

This technique may be difficult to implement as higher voltages than VDD and lower voltages than VSS shall be provided to deeper reverse body biasing (VP and VN respectively). In addition, providing the body potential requires routing the body grid that adds the overall chip area and adds complexity to the chip's routing. Despite the fact that up to three orders of magnitude of leakage current reduction was reported using this technique (Keshavarzi et al. 1999), the effectiveness of reverse body bias decreases with technology downscaling (see section 4.2).

4.1.2.1 Multiple Body Bias

This technique consists on changing the body voltage in bulk silicon devices, in order to modify the threshold voltage. If different body voltages are applied to different transistors, they cannot share the same well; therefore, triple well CMOS technologies are required. However, this technique can be directly implemented in partially-deplected Silicon on Insulator (SOI) transistors, since such devices are naturally isolated.

4.1.2.2 Dynamic Threshold CMOS

In the dynamic threshold CMOS (DTCMOS) technique, the threshold of devices is modified dynamically to suit the operating state of the circuit. This can be achieved by tying the body and the gate of the device together (Assaderaghi et al. 1997). As shown in Fig. 4.7, the bodies of both transistors in the inverter of the example are connected to the input. To do so, a triple-well CMOS technology process is needed for a proper bulk connection; however, stronger advantages of such principle can be achieved by using partially depleted Silicon-on-Insulator (SOI) devices, as each device is perfectly isolated from the others.

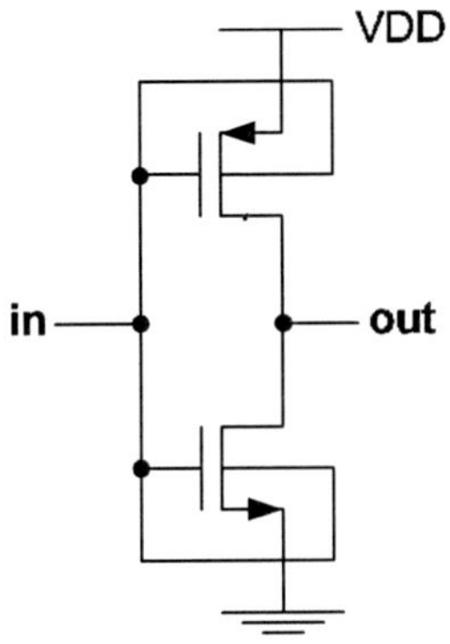

Fig. 4.7 DTCMOS principle: inverter

The supply voltage of DTMOS technique is limited by the diode built-in potential. In fact, the pn junction diode between the source and body should be reverse biased. Therefore, this technique is only suitable for ultra-low-voltage circuits (0.6 V and below), most of the times related to digital implementations.

Dynamic threshold voltage scaling is another technique for active leakage power reduction mostly used in digital systems. This scheme utilizes dynamic adjustment of frequency through back-gate bias control depending on the system's workload: when the workload decreases, less power is consumed by increasing the threshold voltage. Most common dynamic threshold scaling techniques are threshold voltage hoping scheme (Nose et al. 2002) and Dynamic threshold voltage scaling scheme (DVTS) (Iijima et al. 2006).

4.2 Gate Length Downsaling

The strong demand on new electronic devices with better performances and at lower cost has lead the electronic industry to the production of devices in submicron CMOS technologies (Baschirotto 2009) reducing intensively the length of the transistor gate. Due to this reduction, a noticeable increase in maximum transistor switching frequency is obtained and, as a consequence, there is an enhancement of the performances at the same time that power consumption is reduced for microprocessors and digital memories.

Supply voltage downscaling is perhaps the most effective way of reducing logic circuits power (Chen et al. 2010), as the dynamic switching of transistors scales quadratically with VDD. Voltage downscaling also increases latency, especially when VDD is scaled to the subthreshold region. Even though leakage power also scales down with VDD, leakage energy per cycle increases because of this increased latency. However, as shown in Figure 4.2, the competing trends in dynamic and leakage energy result in an intermediate voltage where total energy per cycle is minimized (Zhai et al. 2004).

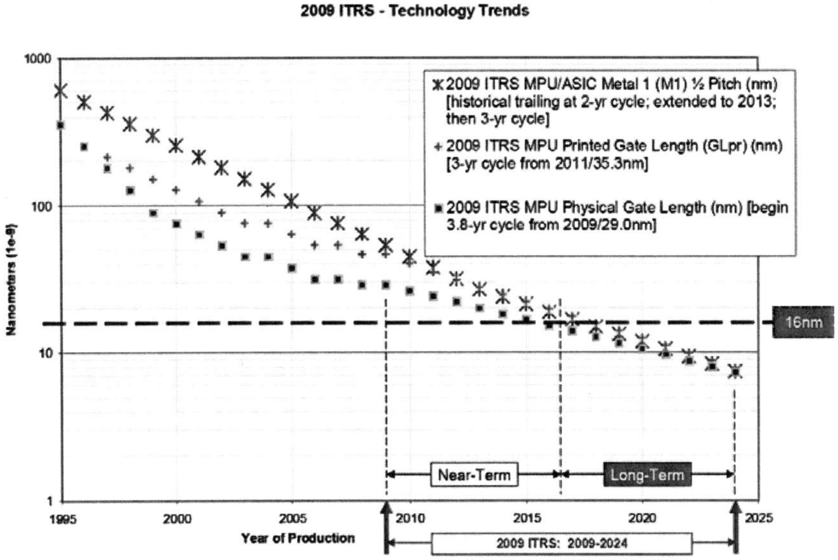

Fig. 4.8 Reduction of the size of MOS transistors in the last years (ITRS Roadmap)

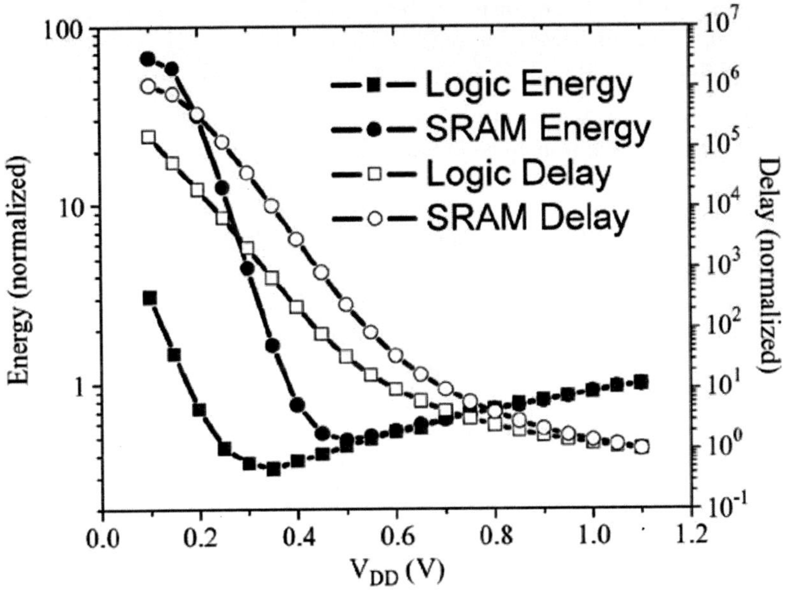

Fig. 4.9 Dynamic energy and leakage power vs. supply voltage (Chen et al. 2010)

However, analog circuits do not benefit from technology scaling, as several second order effects in the transistor that were irrelevant with classical techniques become now dominant, (i.e the increase in the leakage current or the decrease on the output resistance). These effects highlight the need to develop new design techniques based on smart biasing, auto-calibration and/or oversampling, which, at the end, cause an increase in power consumption.

Some of the major challenges that analog designers have to face are:

- Improving signal to noise ratio (SNR). As technology scales down, the available signal swing decreases with the supply voltage. In addition, the threshold voltage does not diminish in the same proportion so that the available voltage headroom is even more limited.
- Achieving a high gain. Precision in analog design is obtained thanks to feedback, which requires a high gain to be effective. The small signal gain of the transistor is given by gm/gds, where gm is the transistor transconductance and gds is the output conductance. As the technology scales down, the transistor gain of the basic MOS transistor decreases significantly from typical 15.2 to 6.1 for a 250 nm and 65n m technology (Pekarik et al. 2004) respectively.
- A high leakage current. In modern technologies, the leakage current through the gate of the MOS transistor and the drain current for a transistor in the cut-off region are no longer negligible.

- Bad matching. The matching between MOS transistors improves with the scale of the technology (Baschirotto 2009), (Pekarik et al. 2004), (Annema et al. 2005); however, small devices suffer from a higher mismatch as it is inversely proportional to $\sqrt{WL}$ (Iwai 2005), (Garget al. 2003). In addition, quantum effects could be limiting the matching as the oxide width has been reduced to only a few layers. In fact, these effects are becoming so important that they are also affecting digital circuits as, for example, SRAM memories (Singh et al. 2009), (Torrens et al. 2010).
- Large deviation of transistor parameters. Several effects, most of them difficult to control, cause large variations in the threshold voltage of the MOS transistors in deep submicron technologies. These effects include short channel effects; the influence of vertical fields on the mobility of charge carriers, lateral field or Drain Induced Barrier Lowering (DIBL). These variations make the dispersion in the analog behavior one of the major problems encountered in nanometer technologies, so that further research is needed to guarantee correct operation of the analog circuits in spite of changes in the device parameters.

On the other hand, technology downscaling also causes significant positive effects such as:

- Maximum operating frequency. The maximum operating frequency of transistors increases significantly.
- MIM capacitors. The characteristics of the vertical Metal-Insulator-Metal capacity benefit from the increase in the number of metals layers in modern technologies and the improvement in the control of photolithography process that allows for capacitors with improved matching and lower parasitic capacitances. A short study of the capacitance values of MIM capacitors by both UMC and TSMC 90 nm is summarized in section 4.1.2 and 4.2.2 respectively.

The highly positive consequences of technology downscaling in the logic circuitry can be used to digitally compensate the lack of precision of analog circuits. Therefore, mixed signal circuit design without using high precision analog blocks is one of the most promising research lines in analog microelectronics (Chae et al. 2008), (Veldhoven et al. 2008), (Fiorenza et al. 2006), (Anthony et al. 2008), (Nazemi et al. 2008). Described below are some current trends in analog microelectronic design for deep submicron technologies:

- Trends at the circuit level:
  - New dynamic biasing techniques (Oh et al. 2009), (Wang et al. 2008), which reduce the power taken from the supply when the input signal allows it.
  - New techniques to compensate for manufacturing process (Qazi et al. 2011), (Mukadam et al. 2010) based on tuning circuits and/or local feedback.

  - Validation of current low power design techniques for nanometer technologies.
  - New techniques to increase the gain in amplifiers mainly based on “gain boosting” and cascade amplifiers (Ahmadi 2006).
- Trends at the system level: Use of digital circuits to compensate for the lack of precision of analog blocks mainly through digital calibration, analog circuit redundancy and/or oversampling (Lewyn et al. 2009).

4.3 Silicon-on-Insulator (SOI)

One of the major issues (if not the most) that designers must deal with when designing ICs is the reduction of the power consumption. As it has been previously mentioned, focus on power has been driven by increases in device leakage as a result of the technology downscaling. In addition, isolation scaling in CMOS processes beyond 32 nm will likely result in an exponential growth in bulk process complexity. Such issues are not present in SOI technology, what has dramatically accelerated the interest in SOI as an alternative to bulk CMOS technologies in the last years.

SOI has many well know advantages over bulk technologies. Improved chip performance due to lower junction capacitance, floating body, and improve short channel effect, significantly better SER immunity and the elimination of latch up as a design concern are all clear SOI advantages. But for much application, improved power will be most interesting to designers, appart from the possibility of implementing design issues that have been on the mind of CMOS bulk design community over the years.

4.3.1 Technology Description

In traditional CMOS bulk technologies transistors are implemented on a p-type semiconductor substrate, as it has been described in section (xx). On the contrary, SOI technology transistors are built on a silicon layer, but resting on an insulating layer (commonly Silicon Dioxide, SiO_2). There are different ways of implementing the insulating layer, e.g. by flowing oxygen onto a plain Silicon wafer and then heating the wafer to oxidize the silicon, thereby creating a uniform buried layer of silicon dioxide. This way, transistors are encapsulated in SiO2 on all sides. The principal benefit of SOI technology is that such insulating layer increases device performance by reducing junction capacitance, as the junction is isolated from bulk silicon. This fact implies great impact in the devices’ input impedance and driving capabilities: higher speed and lower dynamic power consumption are achieved.

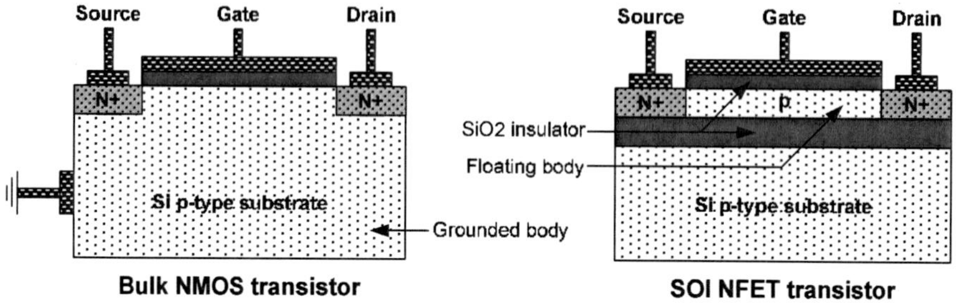

Fig. 4.10 Bulk CMOS and SOI transistor cross section comparison

Shown in Fig. 4.10 is a cross section of a NMOS transistor implemented in both traditional CMOS bulk and SOI technologies respectively. As it can be see, in a standard bulk CMOS process technology, the p-type body of the NMOS transistor is held at the lowest potential of the system (e.g. ground voltage), while the n-type body of a PMOS transistor would be held at the highest potential present in the circuit, commonly supply voltage (VDD). On the other hand in SOI technology process, the source, body and drain regions are physically isolated from the substrate. The body of transistors is typically left unconnected (floating body), what makes it to get freely charged / discharged due to switching activity (e.g. digital electronics), and this condition affects the threshold voltage (see section 4.1) and many other device characteristics.

A direct consequence of the floating body is related to the die area: as there is no need for metal contacts to wells used for traditional bulk CMOS transistors, the overall area occupied by devices in an integrated circuit is reduced.

Depending on the thickness of the silicon-insulating layer, two different SOI technologies arise: partially depleted SOI and fully depleted SOI. They are shortly described in the following paragraphs.

4.3.1.1 Partially Depleted SOI

On the other hand, if the insulated layer of silicon is made thicker, the inversion region does not extend the full depth of the body (). A technology designed to operate this way is called a "partially depleted" SOI technology (PD-SOI). The undepleted portion of the body is not connected to anything. The exact voltage depends on the history of source, gate, and drain voltages leading up to the current time, what is known as the history effect (see section xxx).

However, the voltage can be expected to fall within a known range. The body voltage affects the conduction of the channel and therefore the switching speed and parasitic capacitance of the circuit. In an NMOS transistor, a lower initial body voltage results in a thinner inversion layer, lower conductivity, and slower switching. Conversely, a higher initial body voltage results in faster switching. In a PMOS transistor, the opposite occurs; a lower initial body voltage results in faster switching.

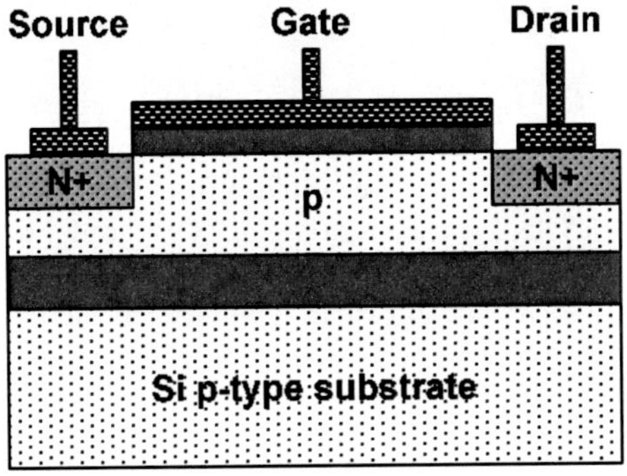

Fig. 4.11 Partialy depleted SOI technology process cross section

Charging of the transistor body (floating body) leads to a change in the threshold voltage (see section 4.1).If this properly is taken into account in the IC design, then the result is a faster switching, thus higher performance at same VDD.

4.3.1.2 Fully Depleted SOI

In an NMOS transistor, applying a positive voltage to the gate depletes the body of p-type carriers and induces an n-type inversion channel on the surface of the body.If the insulated layer of silicon is made very thin, the layer fills the full depth of thebody, avoiding a floating voltage. A technology designed to operate this way is called "fully depleted" SOItechnology (FD-SOI). In 45nm and below CMOS, the threshold voltage can be tuned with a mid-gap metal gate, leaving the fully depleted body undoped.

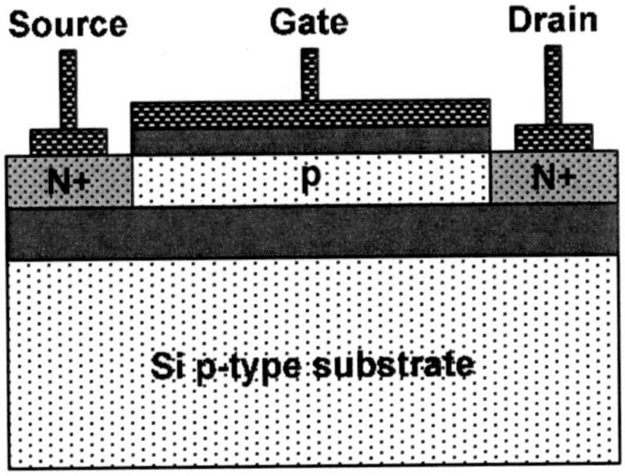

Fig. 4.12 Fully depleted SOI technology process cross section

With such technology, higher channel mobility and therefore higher performance are achieved, as well as lowervariability from one device to another, what is a very important issue in analog circuits. Fully Depleted SOI enables a CMOS low power technology with undoped body, what gives the best performance in terms of low leakage coupling, a perfect choice for Low Power Applications.

4.3.2 SOI Technology Benefits in Analog Circuits

It is evident that SOI CMOS technology offers substantial digital circuit performance improvements over bulk CMOS at the same technology node. These improvements are more evident as technology processes scale down. Therefore, more and more CMOS fabs are transitioning to SOI in order to enable these gains.

As CMOS (firstly intended for digital circuits) has led both analog and RF design trends, now analog designers try to benefit from such performance gains by accommodating analog, mixed-signal and RF systems into SOI-enabled CMOS technologies.

4.3.2.1 Latch-Up Elimination

Among the most important aspects in which SOI differs from bulk are substrate noise isolation, body voltage control, self-heating, and PN junction diode characteristics. The buried oxide layer that insulates SOI devices from the bulk substrate acts as a capacitor dielectric and blocks DC signals from coupling between devices. The most obvious advantage of this is in eliminating the possibility of latchup.

This has the further benefit of eliminating the need for extensive substrate or well contacts, saving die area. The degree to which the buried oxide blocks AC signals depends strongly on the substrate resistance and the frequency.

4.3.2.2 Substrate Noise

With the increased integration of digital and analog circuits on the same chip, substrate noise issue is dominant in CMOS bulk processes. Especially the digital noise can affect the sensitive analog circuits, as represented in Fig. 4.13.

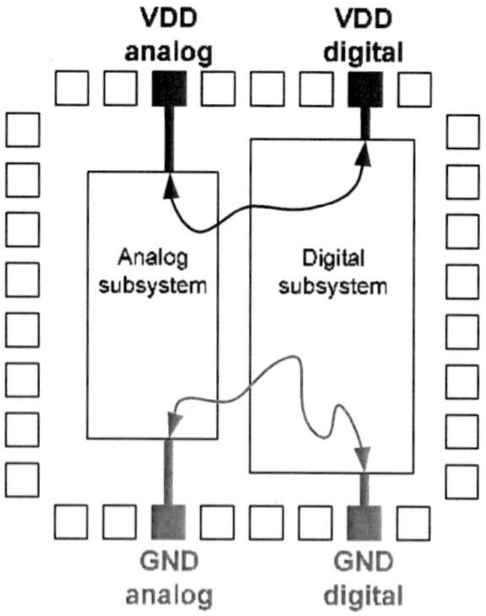

Fig. 4.13 Substrate noise coupling

Although the digital circuitry might have its own VDD and VSS (or GND) pins, as switching signals are present, current spikes through the VDD and VSS lines are produced. These spikes are due to charging and discharging capacitances and short circuit currents as present in CMOS logic gates. Such signals propagate through the ohmic substrate and may couple to the analog supply or ground lines, degrading the performance of analog circuits. On the other hand, in SOI technology the buried oxide layer acts a dielectric barrier and it helps reducing the substrate noise.

However, isolating circuits with the use of SOI technology may seem to be a very good idea, but the isolating layer is not a perfect insulator. The parasitic capacitance of the silicon oxide layer results in that higher frequency components are partially by-passed. Therefore, lower frequencies are effectively attenuated in SOI, while higher frequencies are less attenuated. Therefore, special care must be taken when dealing with high-speed digital signals and RF signals in the same chip.

4.3.3 SOI Design Issues Not Present in CMOS Bulk

There are three main unique SOI circuit design issues due to floating body and the insulation layer of the SOI wafer that CMOS bulk circuit designers do not have to deal with:history effect, bipolar current and local heating.

4.3.3.1 History Effect

The body of the NFET or PFET in the SOI wafer is floating instead of connected to GND (NMOS) or VDD (PMOS) as in bulk CMOS processes. As described above, this floating body can change the FET threshold due to differences in the body voltages. This could cause variation in the circuit delay and mismatch between two identical devices. As the SOI circuit switches, the body voltages of the switching transistors will change from their previous steady state condition. This is called the history effect. This is one of the most interesting circuit design issues in SOI but it is also a benefit which contributes to its performance advantage over bulk CMOS. Learning how to model and predict the body voltage of SOI transistors will lead to a successful SOI circuit design (Jeong-Hyong et al. 2002), (Daghighi and Asgari-Khoshooie 2010).

This effect is particularly interesting in digital circuits, due to their switching character. A SOI logic circuit can have different (faster) delay if switching regularly verses a circuit that has been inactive for a long time and then switches. If a circuit is not active for long enough time to be in a steady state and then switches, this switching activity is called first switch. If the circuit is switching more regularly, this is called second switch. Typically, second-switch has faster delay than first-switch due to the fact that the body to source voltage of the second-switch is higher than first-switch, which lowers the threshold of the second-switch transistor.

- **Digital static CMOS circuit.** There are no real functional concerns when designing such circuits due to history effect: the same circuit designed to work in bulk CMOS can be used for SOI without any modifications. The main difference is in the timing model methodology (Shahriari and Najm 2001), (Valentian et al. 2004).
- **Differential amplifier.** Due to history effect, the input differential pair can have different body voltages and create mismatch between them. To address this type of design issue, SOI technology provides other type of transistors such as the body contact transistors. The body contact transistors have one additional terminal to provide the circuit designers the option of connecting the body to any nodes in their designs. One common usage is to tie the body to GND (NFET) or VDD (PFET) to reduce or eliminate the history effect. In this case, the SOI transistors behave similar to bulk CMOS transistors. The negative effects of the body contact transistors are that they are about two times larger and switching slower than the floating body transistors. Therefore, it is recommended to use body contact transistors only in places that good matching between devices is needed.
- **Other analog designs.** On a device operating in an analog circuit, a body contact may be needed. Without a body contact, the potential of the body will float to a value which is dependant upon the biases applied to the gate, source and drain contacts, and will also become a function of time, impacting output resistance of the device and its threshold-matching to the next device. The electrical behavior of SOI transistors can be significantly different from bulk CMOS transistors because of floating body effects. In typical analog designs, body contact transistor can be used (i.e. current sources or any matching transistors designs) in order to eliminate the floating body effect.

4.3.3.2 Bipolar Current

There is a low gain parasitic bipolar transistor on every floating body SOI FET transistor (Armstrong and French 1992). As depicted in Fig. 4.14, this bipolar transistor is in parallel with the FET transistor and could cause false switching to the off FET transistor.

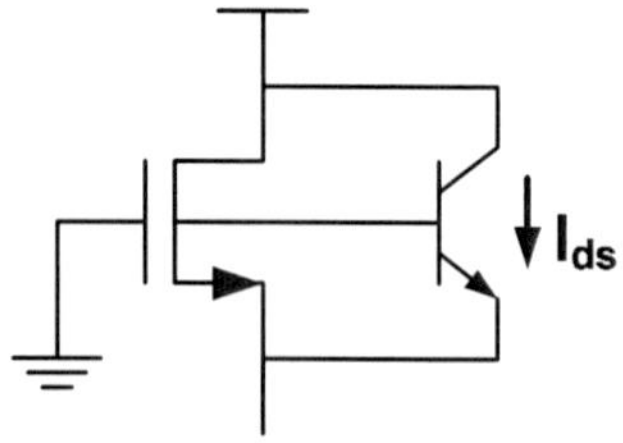

Fig. 4.14 Bipolar current of a SOI FET

However, over the years of the technology downscaling, this bipolar current effect has been pretty much eliminated due to the reduction of the operating voltage of the 90nm technology node and beyond. In some extreme cases, where the design has many transistors connected in parallel, the designer needs to verify the bipolar current to ensure the functionality of the circuit. This is particularly important if the design requires functional burn-in at a much higher voltage than the typical operating voltage condition.

4.3.3.3 Local Heating

The insulation layer of the SOI wafer creates a potential temperature delta between devices called local heating. Circuits with DC current, such as the case of analog (and RF) and mixed-signal circuits, may have local heating effect. However, this can be reduced or eliminated by using robust layout techniques for analog designs which are required for current density and electro migration issues and which are commonly used in bulk CMOS technologies. These requirements also help to spread the temperature of the analog circuits and minimize the temperature delta.

4.3.4 SOI and IC Design for Radio Frequency

When an RF chip is built on a traditional CMOS bulk silicon substrate, the semiconducting properties of the silicon induce RF signal loss in the substrate. These capacitive and resistive losses negatively impact the energy consumption of the circuit. Apart from ohmic losses, semiconducting properties of the silicon also induce transmission of parasitic interferences (i.e. crosstak, see Fig. 4.15). However, the use of a SOI substrate improves significantly the high frequency behaviour of the chip, due to the following reasons:

- First, the buried insulating layer reduces part of the electromagnetic field propagation
- Second, because bonded SOI technology enables the use of a highly resistive (intrinsic silicon) handle wafer, hugely reducing both resistive losses and crosstalk.

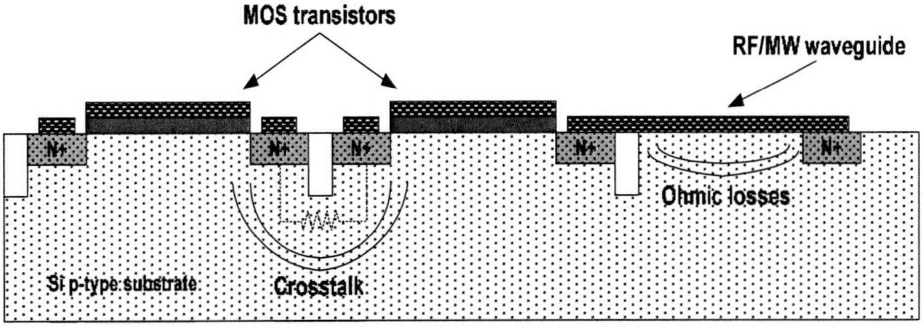

Fig. 4.15 RF circuit on CMOS bulk technology

On the contrary, SOI substrates open new perspectives for RF & SoC circuit designers due to their high-resistivity (HR) characteristic. Functions usually requiring expensive III-V compounds (e.g. antenna switch) can now be integrated on silicon, reducing the overall system cost with comparable performance and a higher integration level. Denser chip layouts are also achievable thanks to insulation improvement (see Fig. 4.16). SOI also enables processed top layer transfer onto electronically inert substrates (e.g. glass), further improving the RF performance.

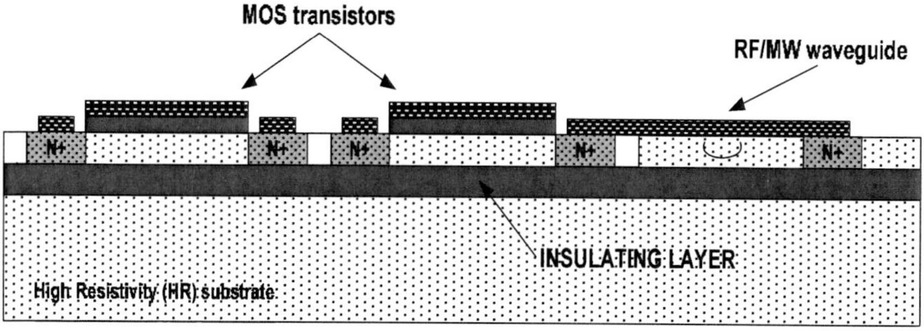

Fig. 4.16 RF circuit on SOI using high resistivity (HR)

As traditional benefits of SOI CMOS technology also include the speed versus power-consumption trade-off, this designates SOI as the ideal platform for low-power RF systems. It is compatible with lateral bipolar transistors integration and with future transistor architectures like FinFETs (Ponton et al. 2009).

References

(Ahmadi 2006) Ahmadi, M.M.: A new modeling and optimization of gain-boosted cascode amplifier for high-speed and low-voltage applications. IEEE Transactions on Circuits and Systems II: Express Briefs 53(3), 169–173 (2006)

(Alvarado et al. 2010) Alvarado, U., et al.: Low Frequency Noise Analysis and Minimization in Gilbert-Cell Based Mixers for Direct Conversion (Zero-IF) Low-Power Front-Ends. International Journal of Circuit Theory and Applications (Wiley) 38, 123–129 (2010)

(Andricciola and Tuinhout 2009) Andricciola, P., Tuinhout, H.: Influence of halo doping profiles on MOS transistor mismatch. In: Annual Workshop on Semiconductor Advances for Future Electronics and Sensors, SAFE (2009)

(Annema et al. 2005) Annema, A.-J., Nauta, B., van Langevelde, R., Tuinhout, H.: Analog Circuits in Ultra-Deep-Submicron CMOS. IEEE JSSC 40(1) (January 2005)

(Anthony et al. 2008) Anthony, M., et al.: A Process-Scalable Low-Power Charge-Domain 13-bit Pipeline ADC. In: Dig. VLSI Circuits Symposium (June 2008)

(Armstrong and French 1992) Armstrong, G.A., French, W.D.: Suppresion of Parasitic Bipolar Effects in Thin-Film SOI Transistors. IEEE Electron Device Letters 13(4), 198 (1992)

(Assaderaghi et al. 1997) Assaderaghi, F., et al.: DynamicThreshold-Voltage MOSFET (DTMOS) for Ultra-Low Voltage VLSI. IEEE Transactions on Electron Devices 44(33), 414 (1997)

(Baschirotto 2009) Baschirotto, A.: Analog Design in scaled technologies. IDESA Advanced Tutorial Series (April 2009)

(Chae et al. 2008) Chae, Y., et al.: A 0.7V 36μW 85dB-DR Audio $\Delta\Sigma$ Modulator Using Class-C Inverter. ISSCC Dig. Techn. Papers, pp. 491–492 (February 2008)

(Chang et al. 1998) Chang, M.H., et al.: A Highly Manufacturable 0.25um Multiple-Vt Dual Gate Oxide CMOS Process for Logic/Embedded IC Foundry Technology. In: Proc. Symp. VLSI Technology, Digest of Technical Papers, pp. 150–151. IEEE Press, Los Alamitos (1998)

(Chen et al. 2010) Chen, G., et al.: Circuit Design Advances for Wireless Sensing Applications. Proceedings of the IEEE 98(11) (November 2010)

(Daghighi and Asgari-Khoshooie 2010) Daghighi, A., Asgari-Khoshooie, A.: A width-dependent body-voltage model to obtain body resistance in PD SOI MOSFET technology. In: International Conference on Advanced Semiconductor Devices & Microsystems, ASDAM (2010)

(Fiorenza et al. 2006) Fiorenza, J.K., et al.: Comparator-Based Switched-Capacitor Circuits for Scaled CMOS Technologies. IEEE J. Solid-State Circuits 41(12), 2658–2668 (2006)

(Garg et al. 2003) Garg, M., Suryagandh, S.S.: Scaling Impact on Analog Performance of Sub-100nm MOS-FETs for Mixed Mode Applications (2003)

(Gosset et al. 2008) Gosset, G., et al.: Very High Efficiency 13.56 MHz RFID Input Stage Voltage Multipliers Based On Ultra Low Power MOS Diodes. In: IEEE International Conference on RFID (2008)

(Iijima et al. 2006) Iijima, M., et al.: Dynamic threshold voltage control for dual supply voltage scheme on PD-SOI. IEICE Electron. Express 3(21), 453–458 (2006)

(Iwai 2005) Iwai, H.: Recent Status on Nano CMOS and Future Direction (2005)

(Jeong-Hyong et al. 2002) Yi, J.-H., et al.: Analytical model for a transient floating body voltage in PD-SOI MOSFETs. In: IEEE International SOI Conference (2002)

(Kawaguchi et al. 1998) Kawaguchi, H., Nose, K., Sakurai, T.: A CMOS Scheme for 0.5V Supply Voltage with Pico-Ampere Standby Current. In: IEEE International Solid State Circuits Conference, pp. 192–193 (1998)

(Keshavarzi et al. 1999) Keshavarzi, A., et al.: Efectiveness of Reverse Body Bias for Low Power CMOS Circuits. In: 8th NASA Symposium on VLSI Design, pp. 231–239 (1999)

(Kiat-Seng and Kaushik 2005) Yeo, K.-S., Roy, K.: Low-Voltage, Low-Power VLSI Subsystems. McGraw-Hill, New York (2005)

(Langen and Huijsing 1998) Langen, K., Huijsing, J.H.: Compact Low-Voltage Power Efficient Operational Amplifier Cellsfor VLSI. IEEE Journal of Solid-State Circuits 33(10), 1483–1496 (1998)

(Lewyn et al. 2009) Lewyn, et al.: Analog Circuit Design in Nanoscale CMOS Technologies. Proceedings of the IEEE 97(10), 1687–1714 (2009)

(Mukadam et al. 2010) Mukadam, M.Y., et al.: Process variation compensation of a 4.6 GHz LNA in 65nm CMOS. In: Proceedings of the 2010 IEEE International Symposium on Circuits and Systems (ISCAS), p. 2490 (2010)

(Mutoh et al. 1996) Mutoh, S., et al.: 1-V Power Supply High-Speed Digital Circuit Technology with Multithreshold-Voltage CMOS. IEEE Journal of Solid State Circuits 30(8), 847–854 (1996)

(Narendra et al. 2002) Narendra, S., et al.: Full-chip sub-threshold leakage power prediction model for sub-0.18 μm CMOS. In: Proceedings of the 2002 International Symposium on Low Power Electronics and Design, ISLPED (2002)

(Nazemi et al. 2008) Nazemi, A., et al.: A 10.3GS/s 6bit (5.1 ENOB at Nyquist) Time-Interleaved/Pipelined ADC Using Open-Loop Amplifiers and Digital Calibration in 90nm CMOS. In: Dig. VLSI Circuits Symposium (June 2008)

(Nose et al. 2002) Nose, K., et al.: V_{TH}-Hopping Schemeto Reduce Subthreshold Leakage for Low-Power Processors. IEEE Journal of Solid State Circuits 37(3), 413 (2002)

(Oh et al. 2009) Oh, I.Y., et al.: Synamic bias circuit of CMOS power amplifier sensing dissipated signals. Electronics Letters 45(14) (2009)

(Oowaki et al. 1998) Oowaki, Y., et al.: A Sub-0.1um Circuit Design with Substrate-over-Biasing. ISSCC Digest Technical Papers, 88–89 (1998)

(Pekarik et al. 2004) Pekarik, J., et al.: RFCMOS Technology from 0.25um to 65nm: The State of the Art. In: Proceedings of CICC 2004 (2004)

(Ponton et al. 2009) Ponton, D., et al.: Design of Ultra-Wideband Low-Noise Amplifiers in 45-nm CMOS Technology: Comparison Between Planar Bulk and SOI FinFET Devices. IEEE Transactions on Circuits and Systems I: Regular Papers 56(5), 920–932 (2009)

(Qazy et al. 2010) Quazy, M., et al.: A 512kb 8T SRAM Macro Operating Down to 0.57 V With an AC-Coupled Sense Amplifier and Embedded Data-Retention-Voltage Sensor in 45 nm SOI CMOS. IEEE Journal of Solid State Circuits 46(1), 85 (2011)

(Shahriari and Najm 2001) Shahriari, M., Najm, F.N.: A gate-level timing model for SOI circuits. In: IEEE International Coference on Electronics, Circuits and Systems, ICECS (2001)

(Singh et al. 2009) Singh, T., et al.: Feedback Biasing in Nanoscale CMOS Technologies. IEEE TCASII 56(5) (May 2009)

(Torrens et al. 2010) Torrens, G., et al.: Design Hardening of Nanometer SRAMs Through Transistor Width Modulation and Multi-Vt Combination. IEEE Trans. on Circuits and Systems II 57(4) (April 2010)

(Valentian et al. 2004) Valentian, A., et al.: Modeling subthreshold SOI logic for static timing analysis. IEEE Transactions on Very Large Scale Integration (VLSI) Circuits 12(6), 662–669 (2004)

(Veldhoven et al. 2008) van Veldhoven, R.H.M., et al.: An Inverter-Based Hybrid $\Sigma\Delta$ Modulator. In: ISSCC Dig.Techn. Papers, pp. 493–494 (February 2008)

(Zhai et al. 2004) Zhai, B., et al.: Theoretical and practical limits of dynamic voltage scaling. In: Proc. 41st Annual Design Autom. Conference (June 2004)

(Zhiyu and Kursun 2008) Liu, Z., Kursun, V.: Characterization of wake-up delay versus sleep mode power consumption and sleep/active mode transition energy overhead tradeoffs in MTCMOS circuits. In: Midwest Symposium on Circuits and Systems, MWSCAS (2008)

5 Schematic Design Techniques for Power Saving in RF

RF analog circuits are characterised by the use of relatively simple schemes. This is due to the fact that each component adds some kind of parasitic effects, therefore the design always follows the rule of “less is more”. For this reason, most of power saving techniques widely applied in base band or digital circuits is not suitable for high frequency applications. Chapter 5 outlines the most useful alternatives. In this chapter the general principles are presented, and many references to other works are also included. In order to help the reader to fully understand the impact of these techniques, basic simulations and comparisons are included. The practical application of these techniques is presented in subsequent chapters.

Section 5.1 starts by describing one of the design techniques that has been widely utilised within the examples presented in this book: the current reuse principle. On the other hand, section 5.2 deals with the multi-VDD principle. Sections 5.3 and 5.4 are devoted to the power gating and multiple channel length techniques respectively. Finaly, gate biasing is discussed in section 5.5.

5.1 Current Reuse

Current reuse was one of the first techniques implemented in analog circuits. The benefits are clear, and it has been succesfully applied to many buiding blocks, specially amplifiers and mixers. In the next paragraphs we present the development of the basic idea, followed by some examples.

5.1.1 Operation Principle

The basic idea within this technique is to maintain the transconductance of a NMOS transistor while reducing its drain current to the half (Karanicolas1996). Thus, the power consumption of a circuit can be reduced with no compromise to linearity, gain and other parameters directly related to the transconductance. This principle is illustrated in Fig. 5.1:

Fig. 5.1(a) show a single NMOS device (M1) that has an aspect ratio W/L with a drain current I. With these dimensions, the transconductance of M1 is g_m. On the other side, depicted in Fig. 5.1(b) are two NMOS transistors connected in parallel (M2, M3). Their aspect ratio is the half of the one of M1, and hence half the current of M1 flows through each of them (I/2), but the total current flowing through the compound is again I. In this case, the equivalent transconductance of the compound is $g_{m2,3}=g_{m2}+g_{m3}=g_m$. Therefore, both configurations of Fig. 5.1(a) and Fig. 5.1(b) are equivalent.

U. Alvarado et al.: Low Power RF Circuit Design in CMOS Tech., LNEE 104, pp. 61–85.
springerlink.com

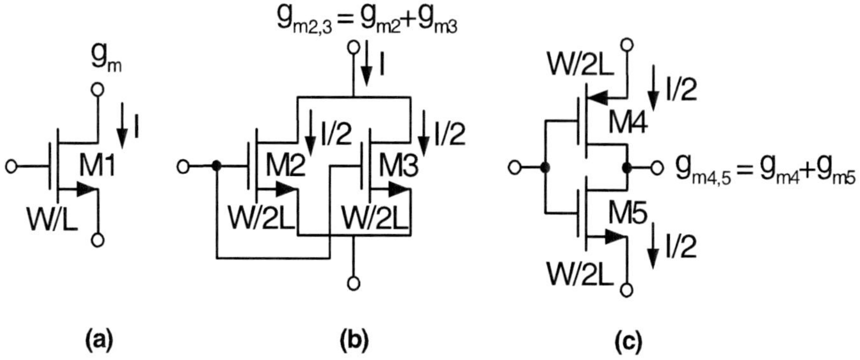

Fig. 5.1 Current reuse operation principle

In Fig. 5.1(c) one of the NMOS transistors has been substituted by a PMOS one (device M4). Both transistors M4 and M5 have the same dimensions of those of Fig. 5.1(b). The equivalent transconductance of the compound is the same as in the previous cases $g_{m4,5}=g_{m4}+g_{m5}=g_{m2,3}=g_m$, but in this case the total current is the half, as the same current flows through both transistors. The input capacitance is nearly equal to that of M1.

In order to study the effect of the configuration of Fig. 5.1 in the thermal noise, let us examine the maximum noise voltage that a single NMOS can generate. Such as maximum output noise occurs when the transistor sees only its own output impedance as the load; that is the case of a single NMOS loaded with an ideal current source, as shown in Fig. 5.2:

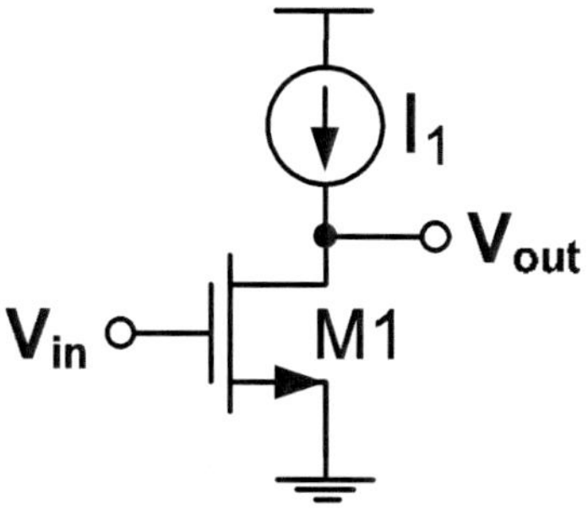

Fig. 5.2 Single NMOS loaded with an ideal current source

The output noise voltage is given by (Razavi 2001)

$$\overline{V_{n,out}^2} = 4kT\gamma r_O^2 \qquad \text{Eq. 5.1}$$

given a noise current (see Fig. 5.3) of

$$\overline{I_{n1}^2} = 4kT\gamma g_m \qquad \text{Eq. 5.2}$$

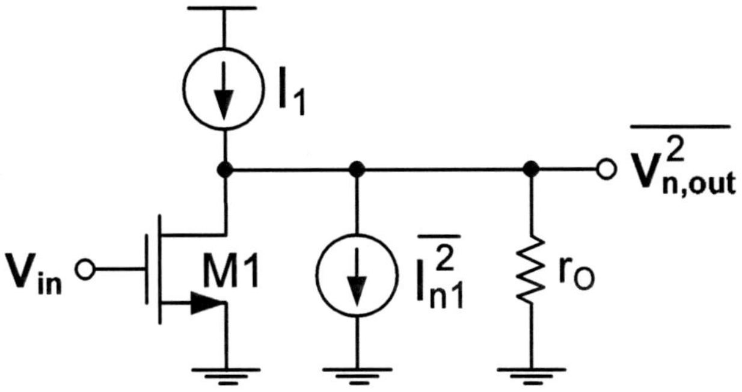

Fig. 5.3 Single NMOS loaded with an ideal current source (noise model)

where $k = 1.38 \times 10^{-23}$J/K is the Boltzman constant, T is the absolute temperature, the coefficient γ is derived to be 2/3 in long channel transistors and may need to be replaced by a larger value for deep-submicron devices (Abidi 1986). In order to have just a qualitative point of view, 2/3 will be used in this analysis.

As the voltage gain for such a basic configuration is equal to $g_m r_O$, then the input referred voltage noise can be written as

$$\overline{V_{n,in}^2} = 4kT \frac{2}{3} \frac{1}{g_m} \qquad \text{Eq. 5.3}$$

Now let us consider the input-referred thermal noise voltage of the current reuse configuration of transistors M1 and M2 depicted in Fig. 5.4:

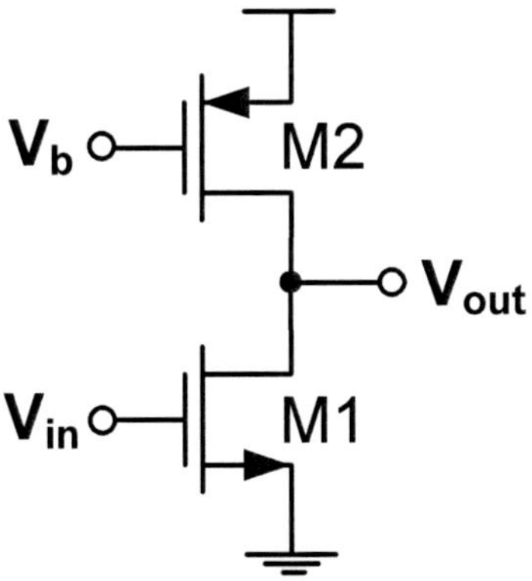

Fig. 5.4 CMOS single pair with current reuse

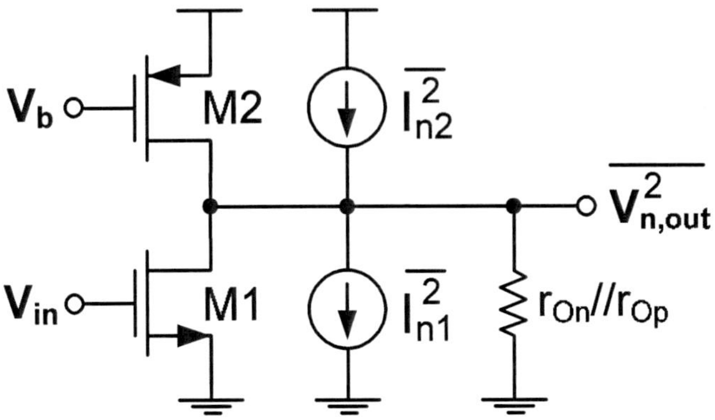

Fig. 5.5 CMOS single pair with current reuse (noise model)

And being its noise equivalent model the one depicted in Fig. 5.5.In this case, the output noise voltage is given by

$$\overline{V_{n,out}^2} = 4kT\left(\frac{2}{3}g_{mn} + \frac{2}{3}g_{mp}\right)\left(r_{On} // r_{Op}\right)^2 \qquad \text{Eq. 5.4}$$

where both g_{mn} and g_{mp} are the transconductances of NMOS (M1) and PMOS (M2) devices respectively, and r_{On} and r_{Op} their output impedances. Being the voltage gain equal to

$$\frac{V_{out}}{V_{in}} = \left(g_{mn} + g_{mp}\right)\left(r_{On} // r_{Op}\right) \qquad \text{Eq. 5.5}$$

The input-referred noise voltage is

$$\overline{V_{n,in}^2} = 4kT\left(\frac{2}{3}g_{mn} + \frac{2}{3}g_{mp}\right)\frac{1}{\left(g_{mn} + g_{mp}\right)^2} \qquad \text{Eq. 5.6}$$

At this point, we could simplify the expression considering that devices M1 and M2 have equal transconductance ($g_{mn} \equiv g_{mp} \equiv g_m$). In fact, this is not the case of currently used CMOS technologies, but it gives us a simple equation useful as a rule of thumb. Eq.5.6 could be rewritten as

$$\overline{V_{n,in}^2} = 4kT\frac{1}{3}\frac{1}{g_m} \qquad \text{Eq. 5.7}$$

which means that the input-referred noise voltage of the current reuse CMOS compound (Fig. 5.4) can be reduced by a factor of two with regard to the configuration of a single NMOS as depicted in Fig. 5.2.

The assumption used in this section is not exact, but useful for an intuitive explanation of the idea. More quantitative conclusions can be derived on the achievable noise and small signal parameters once the following considerations are taken into account. The cut-off frequencies and the noise coefficients of PMOS and NMOS devices are not the same: since the mobility of the PMOS device is lower, $g_{m4,5}$ will be lower than g_m (see Fig. 5.1) Furthermore, for the same aspect ratio and bias currents, NMOS transistors exhibit higher output resistance, providing more ideal current sources and higher gain in amplifiers. For example, referring again to Fig. 5.1, if $\mu_p \approx 0.5\mu n$, then $g_{m4,5} \approx 0.85 g_m$. This reduction in $g_{m4,5}$ for a design could result in approximately several tenths of dB increase in NF with regard to the ideal scenario where both transconductances are identical, which is tolerable considering that the corresponding drain current is reduced by a factor of two.

In addition, the technology down scaling represents and advantage with regard to the use of complementary devices as the difference in the cuttof frequencies of PMOS and NMOS transistors is tending to decrease while, at the same time, the hot electron effect in noise degradation is being reduced for PMOS devices (Gatta et al. 2001).

5.1.2 Basic Implementations

The current reuse principle was first introduced by Karanicolas in (Karanicolas 1996), where a CMOS low noise amplifier and mixer for RF applications are presented. Due to the good results reported by the author, this technique has become very popular in the implementation of other RF and baseband circuits. As it will be shown in chapters 6, 7 and 8, most of the circuits presented in this book have been implemented using such a low-power technique.

This subsection collects the current reuse principle basic implementation in the most representative RF circuits (LNAs, mixers and VCOS).

5.1.2.1 Current Reuse Principle in LNAs

Depicted in Fig. 5.6 is a basic NMOS LNA (Lee 1998), simply composed of a basic common-source NMOS transistor, where both bias and matching circuits have been omitted for simplicity.

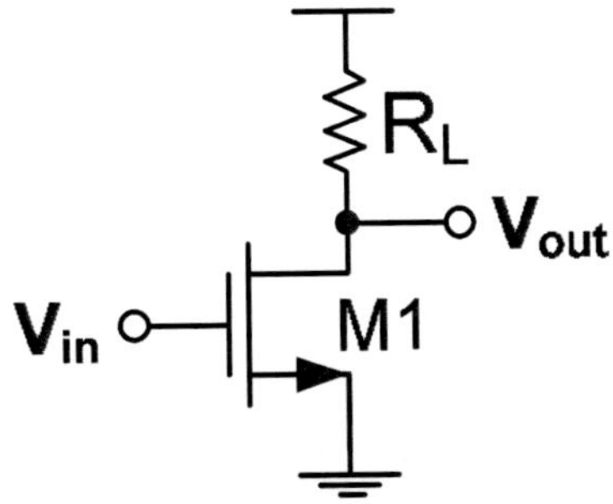

Fig. 5.6 Basic NMOS LNA design

As described in (Karanicolas 1996),nearly same performance of the LNA of Fig. 5.6 but with the half of the power consumption can be obtained by cascading a PMOS device as in the basic current reuse configuration depicted in Fig. 5.4. Such a configuration has been successfully implemented by others (Gatta et al. 2001), (Meng et al. 2007) for low power operation and generally narrowband applications.

Exploiting negative feedback as does the broadband shunt-series amplifier, further studies include a feedback resistor R_F for a wideband input matching performance (Taris et al. 2009), (Hampelet al. 2009), as shown in Fig. 5.7. In this configuration, the absence of a load resistor holds the transistors in saturation region under a minimum supply voltage without compromising design trade-offs. RF remains large to lower the noise figureure., as shown in Eq. 5.8.

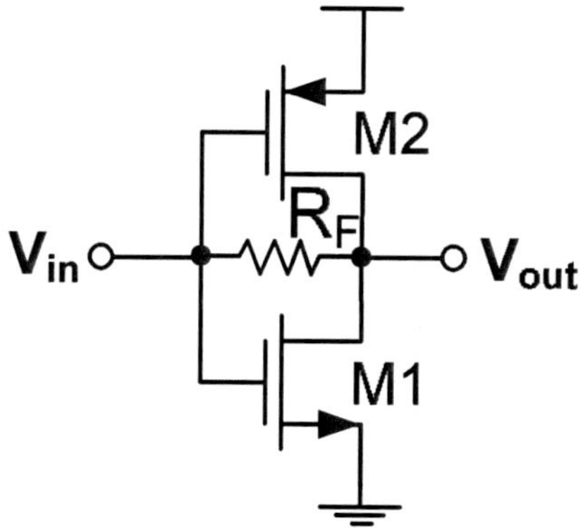

Fig. 5.7 Resistive feedback current reuse LNA configuration

$$NF = 1 + \frac{2}{3}\frac{1}{G_m R_s}\left(\frac{1}{R_s} + \frac{R_s}{R_F^2}\right) + \left(\frac{f}{f_{T\min}}\right)^2 \frac{2}{3} G_m R_s + \frac{R_s}{R_F} \quad \text{Eq. 5.8}$$

where R_s is the source impedance and G_m is the compound transconductance ($g_{m1}+g_{m2}$). On the other hand, the device gain is given by the following expression:

$$V_{gain} \approx G_m \left(R_{load} // R_F\right) \quad \text{Eq. 5.9}$$

where

$$R_{load} = R_F // r_{DS1} // r_{DS2} // R_{out} \quad \text{Eq. 5.10}$$

being r_{DS1} and r_{DS2} the output resistance of devices M1 and M2 respectively, and R_{out} is the input impedance of the subsequent stages seen from V_{out}. Therefore, as demonstrated from Eq 5.8 through Eq 5.10, the inclusion of a large feedback resistor improves the LNA performance in terms of gain and NF, while maintaining a good broadband input matching.

Another key aspect regarding to the low power characteristic of the current reuse technique is that the absence of a load resistor while maintaining devices M1 and M2 in saturation regions allows a low voltage operation, which is a key aspect to minimise the power consumption in deep-submicron CMOS technologies. Eq. 5.11 shows the minimum supply voltage for the current reuse configuration of Fig. 5.7:

$$VDD_{\min} = \sqrt{\frac{2L_p I_D}{\mu_p C_{ox} W_p}} + |V_{Tp}| \qquad \text{Eq. 5.11}$$

where L_p and W_p are the PMOS device dimensions (length and width respectively); μ_p is the hole mobility, and V_{Tp} is the PMOS threshold voltage.

5.1.2.2 Current Reuse Principle in Mixers

The application of the current reuse principle is straightforward in active architectures such as Gilbert-cell based mixers, simply by substituting the NMOS (or PMOS) device of the transconductance stage for a current reuse compound as shown in Fig. 5.4. A simple representation of such Gilbert-cell based active mixerIs shown in Fig. 5.8.

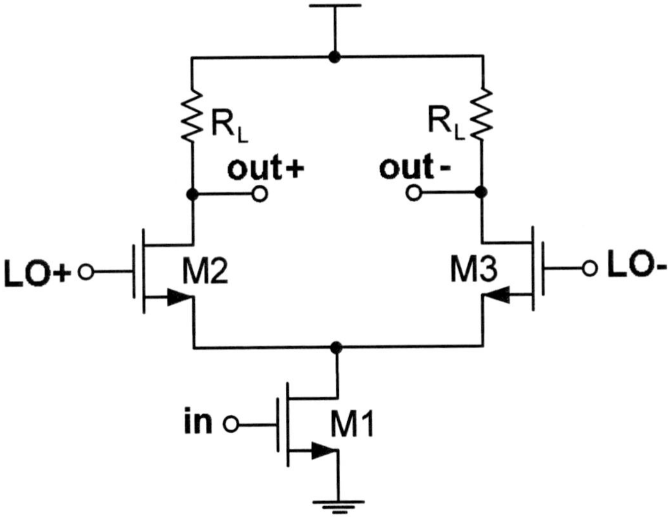

Fig. 5.8 Single-balanced Gilbert-cell based active mixer

where device M1 provides the signal at the input with the voltage-to-current gain for a given transconductance. On the other hand, devices M2 and M3 perform the input signal multiplication with another signal provided by the local oscillator (LO) for frequency conversion. The gain of such a device is given by

$$V_{gain} = \frac{2}{\pi} g_{m1} R_L \quad \text{Eq. 5.12}$$

For a given current flowing through M1, the voltage conversion gain can be doubled (Eq. 5.13) with the insertion of the device M4 in Fig. 5.9, which, along with M1, form part of the current reuse pair.

$$V_{gain} = \frac{2}{\pi}(g_{m1} + g_{m4}) R_L \quad \text{Eq. 5.13}$$

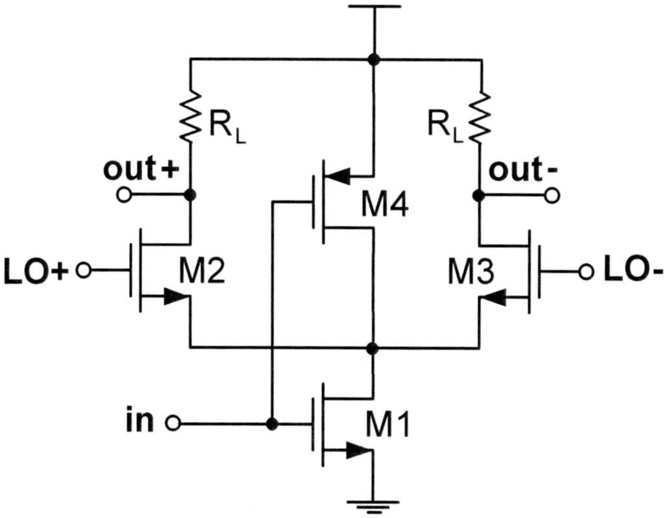

Fig. 5.9 Single-balanced Gilbert-cell based active mixer with current reuse

The configuration of Fig. 5.9 was first introduced by (Lee et al. 2000), providing current bleeding by means of M4 and current reuse, which translates into higher conversion gain, higher linearity, lower noise figureure.and better LO isolation at the output. In addition, the current reduction through the switches (M2 and M3) improves switching efficiency and therefore the low-frequency noise is also reduced (Alvarado et al. 2008), (Alvarado et al. 2010). The noise reduction is briefly described in section 7.3.3 of this book.

Another aspect related to the low-power consumption in active mixers in the low-voltage operation. Current reuse based mixer architectures have been studied folding the switching stage (Vidojkovic et al. 05) in order to allow lower supply voltages. The lowest supply voltage applicable to the circuit of Fig. 5.9 depends on the threshold and overdrive voltages of devices M1 and M4. The overdrive voltages can be written as

$$V_{ov1} = V_{inDC} - V_T$$
$$V_{ov4} = VDD - V_{inDC} - V_T \qquad \text{Eq. 5.14}$$

where V_T is the threshold voltage (typically below 500 mV in 0.18μm CMOS technologies) and V_{inDC} is the gate bias voltage at the gates of M1 and M4. The minimum supply voltage can be therefore calculated as

$$VDD_{\min} = V_{ov1} + V_{ov2} + 2V_T \qquad \text{Eq. 5.15}$$

which indicates that the supply voltage must be higher than 1V. In order to allow a further reduction in the supply voltage, the biasing for M1 and M4 gates can be separated by means of an ac-coupling capacitor, as shown in Fig. 5.10. This way, an ac-coupled complementary current reuse transconductor is obtained by means of the compound M1-M4.

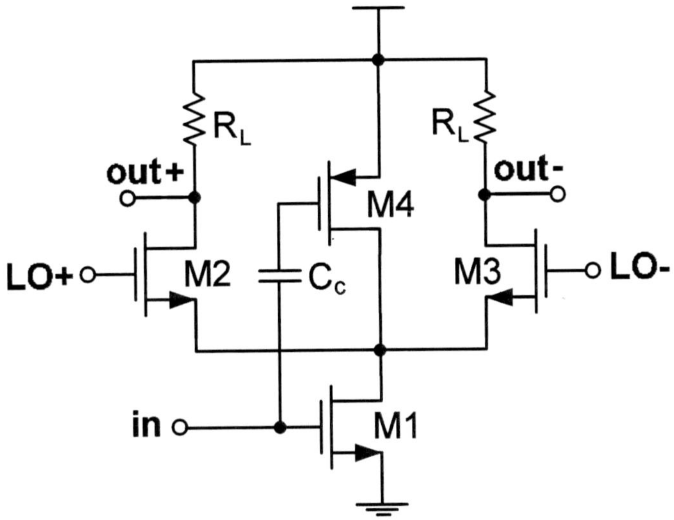

Fig. 5.10 Single-balanced Gilbert-cell based active mixer with current reuse with modified gate biasing

Now the minimum supply voltage becomes

$$VDD_{\min} = V_{ov1} + V_{ov4} + 2V_T + V_{inDC4} - V_{inDC1} \qquad \text{Eq. 5.16}$$

where V_{inDC1} and V_{inDC4} are the biasing voltages at the gates of M1 and M4 respectively. Choosing V_{inDC1} greater than V_{inDC4} the supply voltage can be reduced.

Other architectures exploiting the current reuse design principle have been successfully implemented (Huang et al. 04). However, the Gilbert cell based architecture is the most used configuration for wireless communications as it represents the best trade-off between conversion gain, noise figure, linearity, port isolation and power consumption.

5.1.2.3 Current Reuse Principle in VCOs

Most RF oscillators can be viewed as feedback circuits where, by means of some circuit mechanism, self-sustained oscillation is produced. Theoretically, this phenomenon can be guaranteed with different oscillator architectures (Lee 1998), (Razavi 1998). LC tanks are widely used in RF applications as frequency-selection networks in order to stabilise the output frequency.

However, lossless elements are difficult to realize, and therefore self-sustained oscillation vanishes with time. In the case of LC tank-based oscillators, such a tank, they do not oscillate indefinitely because some energy is dissipated in the parasitic resistance (R_p) in every cycle. If any active network generates a negative impedance equal to or higher than the parasitic resistance ($-R_p$), the total parallel impedance seen by the oscillator will be infinite. Therefore, the energy lost in R_p in every cycle would be replenished by the active circuit.

The most common implementation of RF CMOS oscillators with LC tank uses a cross-coupled NMOS (or PMOS) pair of transistors as active device that provides the circuit with the necessary negative impedance. One possible basic implementation is shown in (Quemada el al. 2009).

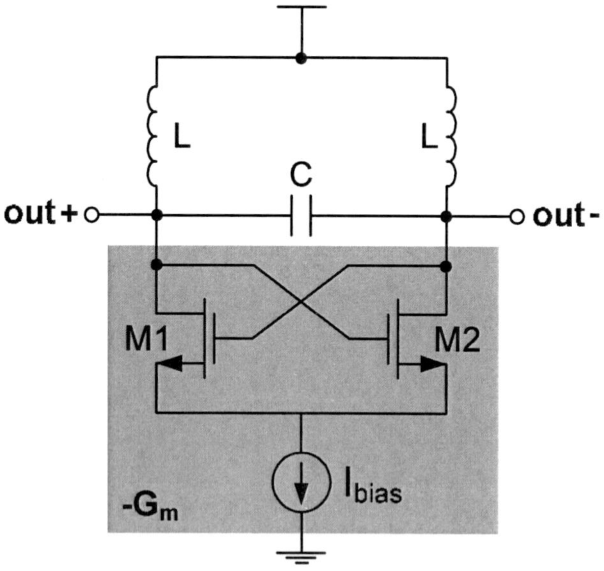

Fig. 5.11 Basic LC tank based NMOS differential oscillator

Considering the cross-coupled feedback oscillator of Fig. 5.11 as a one-port representation, the negative resistance seen at the drain of M1 and M2 can be computed as (Ho et al. 2003):

$$-G_{m(NMOS)} = \frac{g_m}{2} \qquad \text{Eq. 5.17}$$

where G_m is the transconductance of the cross-coupled pair and g_m is the transconductance of each NMOS. In order to guarantee the oscillator startup, it is recommended that the negative transconductance is at least two or three times the loss resistance (Craninckx 1998).

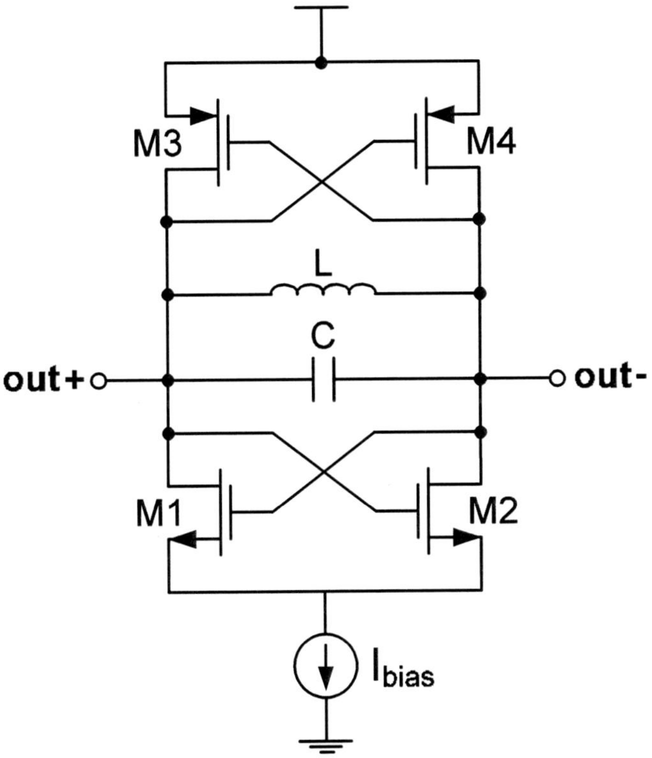

Fig. 5.12 Basic LC tank based CMOS differential oscillator

The CMOS oscillator circuit employs both NMOS and PMOS cross-coupled pairs (Razavi2001) (Fig. 5.12). In a simple CMOS -G_m oscillator the same bias current flows through both he NMOS and PMOS devices, consequently for the same power consumption the configuration yields a negative resistance twice as large (Eq. 5.19). Put another way, the CMOS cross-coupled pairs represent themselves a current reuse configuration, and therefore the same negative transconductance of a single NMOS (or PMOS) cross-coupled pair can be achieved but with half current consumption:

$$-G_{m(CMOS)} = g_m \quad \text{Eq. 5.18}$$

$$G_{m(CMOS)} = 2G_{m(NMOS)} \quad \text{Eq. 5.19}$$

5.1.2.4 Current Reuse Further Considerations

The current reuse principle has been described in the previous subsections for most representative RF building blocks of any wireless front-end for communications. However, the same principle can by applied to other analog circuits, such as baseband OTAs (Chandrashekar and B. Bakkaloglu 2009),mixed-signal systems such as ADCs (Ryu et al. 2007), or whatever circuit as a mean:

- to boost its transconductance for a given current consumption
- to reduce its current consumption for a given transconductance.

Further current reuse RF configurations include merging LNAs and mixers (Sjöland et al. 2003) or even complete front-ends, including LNA, I/Q mixers and VCO (Liscidini et al. 2006),(Cheng et al. 2009), where a single low-power circuit is used for LO generation, low noise amplification and I/Q frequency downconversion. Whereas these circuits obtain a relatively poor performance regarding to the linearity, they represent a very interesting trade-of with regard to gain, noise figureure.and power consumption.

5.2 Multi-VDD

The use of multiple supply voltages (or multi-VDD) is a common technique widely used in digital systems since its introduction during the 1990s (Chandrakasan et al.1992). The first implementations were carried out in audio and video processing circuits, taking advantage of the different needs of computation resources typically found in these applications (KursunandFriedman2006), (Okano et al.2006). The basic idea is the use of two (or more) supply levels: a high value for critical transistors in terms of performance, and a low value for the rest (Sylvester andKaul2001). This technique is especially useful when combined with multi-V_{TH} transistors (Tawfik and Kursun2007), (Dhillon et al.2003). In any case, it is apparent that in order to obtain the maximum power reduction, the power management strategy will be crucial. Several works can be found presenting different approaches, from dual line voltage-island design methods (Wong2007) to more complex algorithms (Cheng et al.2006), (Huaizhiand Wong2007), (Huaizhi et al.2009), (Shibata et al.2010), (Ali et al. 2008),(Gandhi and Mahapatra 2006), (Karimi and Sarrafzadeh 2010), (Sengupta and Saleh 2007), (Khan et al. 2006).

Depicted in Fig. 5.13 is a simplified overview of the multi-VDD technique, where up to n different networks (islands) are supplied with different voltages (VDD_1, VDD_2, …,VDD_n).

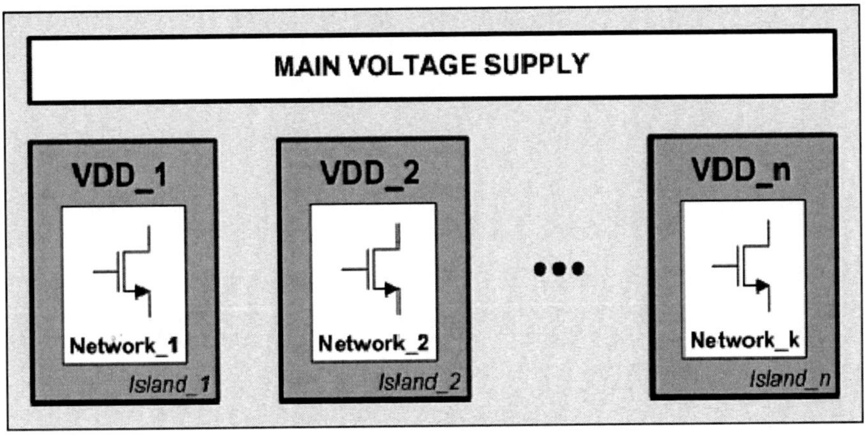

Fig. 5.13 Multi-VDD technique overview

The main drawback of Multi-VDD techniques is the inclusion of an additional level of complexity to the circuit conception (Hobbs and Williams 2008). At system level, power management strategies must be implemented with their corresponding logic circuits. From the architectural point of view, the most significant power consumption penalty comes from the DC-DC converters needed to implement the different voltage supply lines. Many references can be found in this topic, most of them oriented to digital applications (Kulkarni and Sylvester 2003), (Wang et al. 2006), (Ghai et al.2008).

Despite the wide interest shown in digital applications, this technique is not fully developed for analog and mixed mode applications. However, the increasing need for mobile communications devices is driving the quest for methodologies to optimize power consumption at any level. Power amplifier has been one the first building blocks to include dynamic voltage supply. In 2006, I.A. Rippke et al. reported a variable supply voltage power amplifier, analyzing the circuit performance between 1.2V and 2.7V (Rippke et al. 2006). (Chung et al. 2009) proposes an asymmetric multilevel out phasing architecture to control the tradeoff between efficiency and linearity. In (Lee et al. 2006)a single package power amplifier module with a GaAs PA and a Si DC/DC converter for WCDMA application is presented. In this work, a DC/DC converter controls the supply voltage of the PA, increasing the efficiency when the ACPR is not critical. Another field of interest is the amplification of WCDMA output signals. In this case, the inherent high peak-to-average signal ratio limits the efficiency of output stages. J.S. Walling et al. propose an innovative multiple supply (Class-G) linear modulator and PA, which realizes efficient envelope restoration with an appropriate use of the required supply voltage. Finally, a 2.4 GHz PA for WLAN with a supply range between 2.3V to 5.5V can be found in (Li et al. 2010).

Regarding to the complete fron-tend, currently there are no reported works of a complete circuit taking advantage of this concept. However, in (Hsieh et al. 2009) a study of the performance of a fron-tend for low voltage supply is presented. Although this work is not actually focused on Multi-VDD techniques, it shows the methodology that should be applied to develop an appropriate strategy. This is a similar approach as previously shown by F. Khalek et al. in their analysis of a VCO (Khalek et al. 2007).

5.3 Power Gating

Power gating is one of the most effective techniques to optimize power consumption in electronic systems. The basic idea is fairly simple: Some dedicated transistors are used as switches, shutting off power supplies of stand-by circuits. The concept has been applied to digital circuits (Mutoh et al. 1995), (Makino et al. 1999), (Hailin et al. 2005) as it is shown in Fig. 5.14.

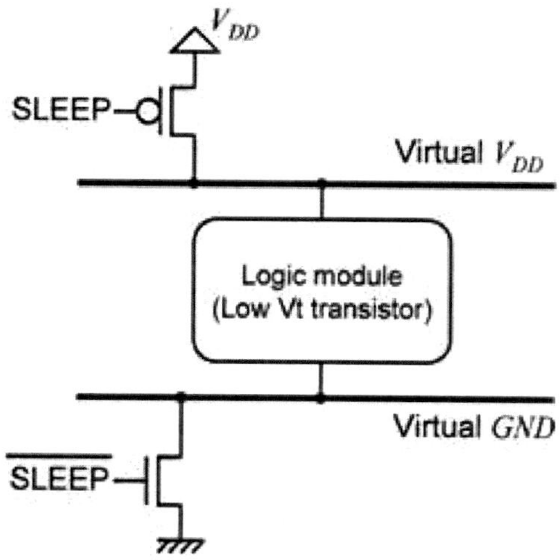

Fig. 5.14 Circuit concept of MTCMOS

The concept is reinforced with the use of Multiple-Threshold Transistors (MTCMOS) (Ping et al. 2010). Logic module can consist of low-threshold voltage transistors for the purpose of high-speed operation. Their power terminals are connected to power sources through high-threshold voltage transistors. Bulk nodes of the module are directly connected to the power supply.

In the active state, SLEEP signal is set to low, and then internal power rails represented with Virtual VDD and Virtual GND function as real power lines. Therefore, the core logic module operates normally at a high speed.

In the off state, SLEEP signal is set high. Virtual lines are floating. Large leakage current of the module, which uses low threshold transistors, is relatively suppressed with the high threshold of the sleep transistors. Therefore, the power consumption during the off state can be drastically reduced by this sleep control.

According to the reports, MTCMOS requires two-side transistors: header transistor connected to the power supply node and footer transistor connected to the ground level. Power gating technique has been proposed instead of MTCMOS and this technique requires only one-side transistor, header or footer. Comparing between MTCMOS and Power gating, the later needs smaller area of cut-off transistor. For the suppression of the off state current, recent SoCs prefer Power gating technique to MTCMOS due to the area occupation.

In cases where a very low V_{DD} is required, a variation of this technique using low threshold transistor, named SC CMOS, has been proposed (Kawaguchi et al. 2000).

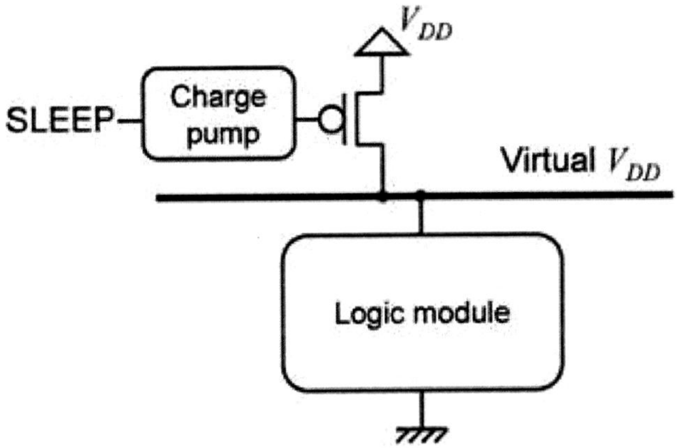

Fig. 5.15 Circuit concept of SC CMOS

Here, a low threshold transistor is employed to cut off the leakage current in the off state. SLEEP signal is boosted through the charge pump circuit and applied to the gate of the cut-off transistor. The applied bias voltage is a little bit higher (0.4 V in the reported study) than the nominal operating voltage of the connected module. A recent study has proposed the optimization of this boosted voltage (Valentian and Beigne 2008).

Bulk Bias Control is another interesting variation to prevent the use of big switches. It relies on the modulation of the threshold voltage to control the leakage current of the transistors. In order to modulate the threshold voltage, a couple of bias control technique has been proposed. These techniques are categorized into two: back bias control and forward bias control In general, back bias is called Reverse Body Bias (RBB) and this one contributes to lower the leakage current. Therefore we will only describe this technique (Nose et al. 2002).

In RBB, the bulk node bias of NMOS transistor is lower than the ground level, and the one of the PMOS transistor is higher than the operating voltage. The effective threshold voltage is higher than the zero-substrate bias threshold voltage. Thus, the leakage current is reduced in the off state with RBB. This principle is briefly described in Fig. 5.16.

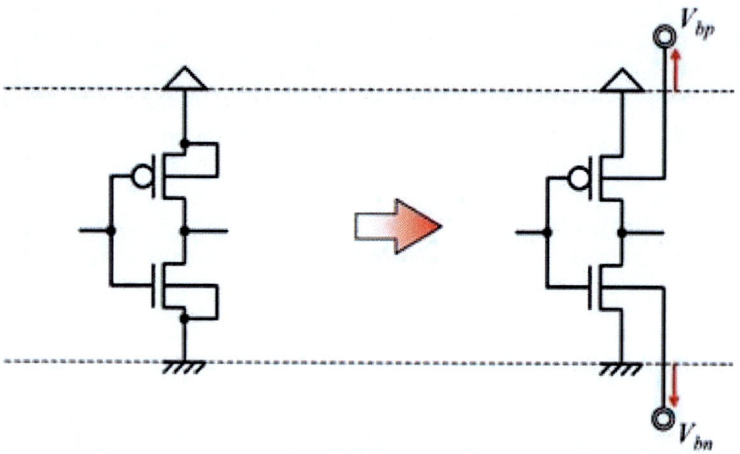

Fig. 5.16 Reverse body bias (RBB)

In order to obtain the bias voltage, a bias generator circuit is required, like charge pump or external dc-dc converter. To calculate the whole power consumption in the off state, the power consumption of the bias generator has to be taken into account.

In any case, the implementation of an appropriate strategy to carry out the power management is crucial. Recent works include adaptive power control algorithms, leading to power reductions up to 12% (Hsieh and Hwang 2010), (Lee et al. 2008).

Regarding analog implementations, power gating has been successfully applied to wireless sensor networks. Panic et al. have demonstrated the benefits of an appropriate gating strategy in low duty WSN (Panic et al. 2008), (Panic et al. 2008(2)). Among the negative effects, the surge current during the wakeup process is one of the most important. (Lee et al. 2009) proposes an innovative wakeup scheduling formulation which considers the trade-off between wakeup times and hardware resources. Finally, the work carried out by (Pasha et al. 2010) shows the importance of power gating in large scale WSN.

5.4 Multiple Channel Length

Scaling-down of transistors size has been a constant trend throughout the last decades. Despite the apparent benefits achieved by small sized devices, there are also some undesirable side effects. For example, an increase in subthreshold leakage current. This problem may be solved using higher threshold transistors in non-critical signal paths (Sirisantana et al. 2000). In cases where multiple threshold transistors are not available, multiple channel length offers a solution.

Due to increase in propagation delay, a careful selection of higher length devices must be observed (Sirisantana et al. 04). In (Gupta et al. 2006) a complete discussion of advantages and drawbacks is presented. There are many methodologies proposed to accomplish this task, oriented to digital system design (FengGao and Hayes 2005). Bol et al. present a reduction of total energy consumption by a factor 5 for a 32nm-node technology (Bol et al. 2005).

5.5 Gate Biasing

The biasing of the transistor becomes an important issue in the analogue design as the technology and the supply voltage are scaled down. Moreover, this polarization presents interesting relations with the drain current of the transistors. The CMOS working regions are analysed in the two first subsections in pursuit of lower power consumption.

5.5.1 *Strong Inversion*

The technology down scaling provides devices with channel widths that are factors of hundreds or even thousands greater than the channel length. These high W/L ratios can result from either wide channel devices, designed to achieve high transconductance, or short channel devices used in high frequency circuits. The analysis of the drain current of a common source transistor in the triode region leads to Eq. 5.20(Lee 1998):

$$I_D = \frac{\mu C_{ox} W}{L}\left[\left(V_{gs} - V_{th}\right)V_{ds} - \frac{V_{ds}^2}{2}\right] \qquad \text{Eq. 5.20}$$

where V_{gs} is the applied gate to source voltage and V_{ds} is the resulting drain to source voltage. (V_{gs}-V_{th}) is normally named overdrive, or V_{eff}. The voltage V_{ds} is $V_{saturation}$ in the strong inversion mode, and this leads to the simplification of Eq. 5.21:

$$I_D = \frac{\mu C_{ox} W}{2L}\left(V_{gs} - V_{th}\right)^2 = \frac{1}{2} g_m V_{eff} \qquad \text{Eq. 5.21}$$

Consequently, the resulting transconductance for this operating region is presented in Eq. 5.22:

$$g_m = \sqrt{\frac{2\mu C_{ox} W}{L} I_D} \qquad \text{Eq. 5.22}$$

Since the drain current is proportional to the increasing of the W/L ratio, a required current for an analogue design may be attained with a lower (V_{gs}-V_{th})

value. The decreasing of the overdrive often establishes the operation of the device in the moderate or even weak inversion. And, as it is outlined in the next paragraphs, the operation below the strong inversion region can result in advantages. More specific expressions are then presented.

As a rule of thumb, these peculiar working regions are delimited in Fig. 5.17, where a numerical approximation of these boundaries is(Comer and Comer 2004):

- Weak inversion $V_{GS} < V_{TH}$ - 80mV
- Moderate inversion V_{TH} +- 80mV < $V_{GS} < V_{TH}$ - 80mV
- Strong inversion $V_{GS} > V_{TH}$ + 80mV

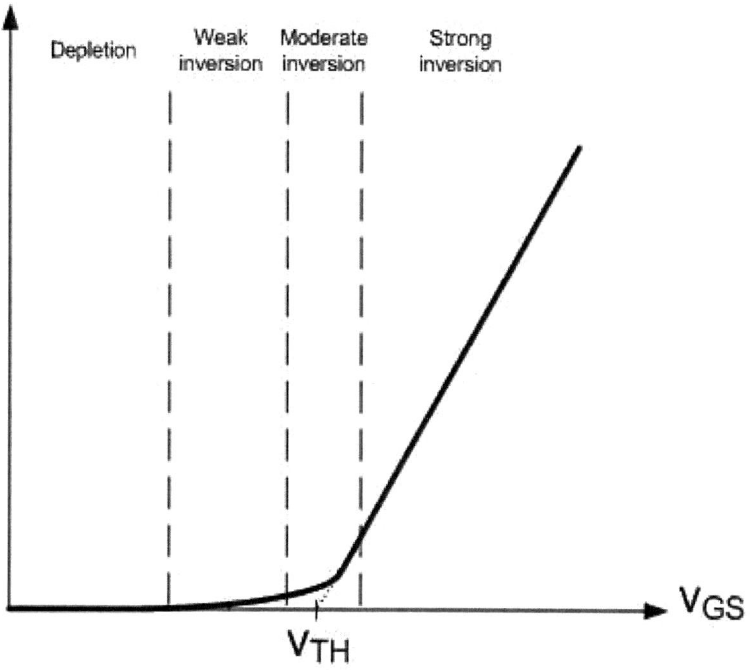

Fig. 5.17 Working regions delimitation in the $I_D - V_{GS}$ plot

Considering the power consumption of the transistors, the currents that limit these regions are definitively more valuable for this analysis. As for the gate to source voltages, these Fig.s are a good estimation (Harrison and Charles 2003), (Tsividis 1997):

- Weak inversion $I_D < 0.1$ Is
- Moderate inversion $0.1\ I_S < I_D < 10\ I_S$
- Strong inversion $I_D < 10$Is

where I_S is called the moderate in version characteristic current and is given by Eq. 5.23:

$$I_S = \frac{2\mu C'_{ox} U_T^2}{\kappa} \frac{W}{L} \qquad \text{Eq. 5.23}$$

Typical values of I_S range from 100nA to 500nA for nMOS with W/L = 1, and from 40nA to 120nA for pMOS with W/L = 1. Of course, for large W/L ratios, the weak inversion region can extend well into the microamp range.

5.5.2 *Weak Inversion*

The weak inversion region is active when the V_{gs} is slightly above the threshold voltage of the transistor. Even negative values of V_{eff} lead to a positive drain current which is called subthreshold current. This unexpected current is primarily due to diffusion but can no longer be considered negligible for wide channel devices. In this region, the simplified relation between I_D and V_{GS} changes from square law to exponential, as provides Eq. 5.24(Allen and Holdberg 1987):

$$I_D = \frac{W}{L} I_{D0} e^{qV_{gs}/nKT} \qquad \text{Eq. 5.24}$$

where I_{D0} is the zero bias current which is process dependant and is related to the voltage from the source to the bulk of the transistor and to V_{th}. q is the elementary charge, n is the subthreshold slope factor, K is the Boltzman constant and T is the temperature in Kelvin degrees. I_{D0} and n should be extracted from experimental data, but n is easily controlled as the minimum value of C_g/C_{ox}(Allen and Holdberg 1987), (Comer and Comer 2004), (Vittoz and Fellrath 1977). Furthermore, Eq. 5.25 displays the resulting transconductance:

$$g_m = \frac{q}{nKT} I_D \qquad \text{Eq. 5.25}$$

This parameter has a clear difference in comparison to the transconductance obtained in the previous subsection: for the weak inversion, it is directly dependant on the drain current whereas it was dependant on the square root of the drain current for the strong inversion. This and other considerations are listed in subsection 5.5.4.

5.5.3 *Moderate Inversion*

This is the last region identified by the models. It's, primarily the transition between the weak and the strong inversion. The change from the exponential formula of the weak inversion region and the quadratic expression of the strong inversion region is not easy to parameterize.

As V_{eff} continues to increase, both the drift current and the diffusion current deserve be considered for the total drain current. And this equilibrium is maintained in the moderate inversion to the point that the drift current dominates the drain current of the transistor (Comer and Comer 2004).

5.5.4 Moderate and Weak Inversion Benefits

The lowering of the overdrive voltage may improve the gain and the harmonic distortion with an even lower power consumption. The transconductance extracted from the model of the weak inversion region presented a direct dependence with I_D, whereas it has been subjected to the square root of the drain current in the strong inversion region. The figure of merit which plots the g_m/I_D ratio depending on the overdrive voltage is shown in Fig. 5.18.The subthreshold region presents the higher and constant g_m/I_D, whereas this FOM decreases as the overdrive voltage is increased over the threshold voltage.

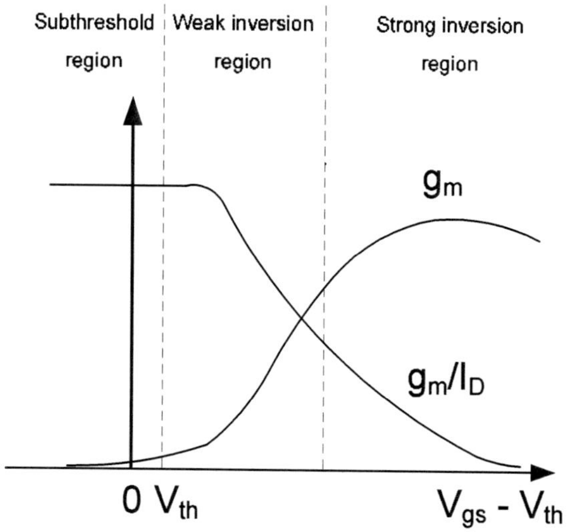

Fig. 5.18 Unit transconductance per current and transconductance depending on the overdrive voltage

Comer et al. outlined that the voltage gain deduced for the two non-strong regions analysed, presented the subthreshold region as beneficial. The weak inversion region had a constant voltage gain. In the strong inversion, the voltage gain decreased as the drain current increased, and consequently the transition (moderate inversion) is also benefited by higher voltage gains.

Three conclusions were experimentally demonstrated in this work:

- Midband voltage gain is maximum in the border between the weak and the strong inversion
- Power dissipation is lower as I_D decreases
- Distortion of the output signal is reduced when I_D is set at the point that maximizes voltage gain.

Nevertheless, this subthreshold biasing manifests a drawback: the operating frequency. This performance is evaluated by Eq. 5.26(Sedra and Smith 2004):

$$f_T = \frac{g_m}{2\pi C_g} \quad \text{Eq. 5.26}$$

Vittoz calculated in (Vittoz and Fellrath 1977) the gate capacitance (C_g) variation with the gate voltage (V_g). This plot performed a minimum around the threshold voltage, while the subthreshold and the strong inversion regions presented a high C_g value. The transconductance is at its lower performance below V_{th}, as depicted in Fig. 5.18. Consequently the maximum available frequency, limited by f_T, is much lower in this region. This restrains the objectives applications of this interesting technique. Vittoz estimates a maximum working frequency of 500MHz in (Vittoz 2003) for a standard CMOS 0.18μm technology.

This drawback, in comparison to the strong inversion region limits the use of the weak and moderate inversion regions to low frequency analogue modules. Depending on the architecture of the transceiver, the most typical use is for post-mixers buffers and moreover baseband operation blocks (active filters, OTAs, etc…)

References

(Abidi 1986) Abidi, A.A.: High-Frequency Noise Measurements on FETs with Small Dimensions. IEEE Transactions on Electron Devices 33, 1801 (1986)

(Ali et al 2008) Ali, N.B.Z., et al.: Delay fault modelling/simulation using VHDL-AMS in multi-Vdd systems. In: 26th International Conference on Microelectronics, MIEL 2008, May 11-14, pp. 413–416 (2008)

(Alvarado et al. 2008) Alvarado, U., et al.: Low Frequency Noise Optimisation in Gilbert-Cell Based Mixers for Direct Conversion (Zero-IF) Receivers. Microwave and Optical Technology Letters 50(12), 3128 (2008)

(Alvarado et al. 2010) Alvarado, U., et al.: Low Frequency Noise Analysis and Minimization in Gilbert-Cell Based Mixers for Direct Conversion (Zero-IF) Low-Power Front-Ends. International Journal of Circuit Theory and Applications 38(2) (March 2010)

(Bol et al. 2008) Bol, D., et al.: Impact of Technology Scaling on Digital Subthreshold Circuits. In: IEEE Computer Society Annual Symposium on VLSI, ISVLSI 2008, April 7-9, pp. 179–184 (2008)

(Chandrashakar and Bakkaloglu 2009) Chandrashakar, K., Bakkaloglu, B.: A 10b 50MS/s Opamp-Sharing Pipeline A/D With Current-Reuse OTAs. In: IEEE Custom Intergrated Circuits Conference, CICC (2009)

(Chandrakasan et el. 1992) Chandrakasan, A.P., et al.: Low-power CMOS digital design. IEEE Journal of Solid-State Circuits 27(4), 473–484 (1992)

(Cheng et al. 2009) Cheng, K.-W., et al.: A 7.2mW Quadrature GPS Receiver in 0.13μm CMOS. In: IEEE International Solid State Circuits Conference, ISSCC (2009)

(Craninckx and Steyaert 1998) Craninckx, J., Steyaert, M.: Wireless CMOS Frequency Synthesizer Design. Kluwer Academic Publishers, Norwell (1998)

(Chen et al. 2006) Chen, D., et al.: Optimality study of resource binding with multi-Vdds. In: 43rd ACM/IEEE Design Automation Conference (2006)

(Chung et al. 2009) Chung, S., et al.: Asymmetric multilevel outphasing architecture for multi-standard transmitters. In: Radio Frequency Integrated Circuits Symposium, RFIC 2009, June 7-9, pp. 237–240. IEEE, Los Alamitos (2009)

(Comer and Comer 2004) Comer, D.J., Comer, D.T.: Operation of analog MOS circuits in the weak or moderate inversion region. IEEE Transactions on Education 47(4) (2004)

(Dhillon et al. 2003) Dhillon, Y.S., et al.: Algorithm for achieving minimum energy consumption in CMOS circuits using multiple supply and threshold voltages at the module level. In: International Conference on Computer Aided Design, ICCAD 2003, November 9-13, pp. 693–700 (2003)

(FengGao and Hayes 2005) Feng, G., Hayes, J.P.: Total power reduction in CMOS circuits via gate sizing and multiple threshold voltages. In: Proceedings of 42nd Design Automation Conference, June 13-17, pp. 31–36 (2005)

(Gandhi and Mahapatra 2006) Gandhi, K.R., Mahapatra, N.R.: Exploiting Data-Dependent Slack Using Dynamic Multi-VDD to Minimize Energy Consumption in Datapath Circuits. In: Proceedings of Design, Automation and Test in Europe, DATE 2006, March 6-10, pp. 1–6 (2006)

(Gatta et al. 2001) Gatta, F., et al.: A 2dB Noise figureure. 900-MHz Differential CMOS LNA. IEEE Journal of Solid State Citcuits 36, 1444 (2001)

(Ghai et al. 2008) Ghai, D., et al.: A Dual Oxide CMOS Universal Voltage Converter for Power Management in Multi-VDD SoCs. In: 9th International Symposium on Quality Electronic Design, ISQED 2008, March 17-19, pp. 257–260 (2008)

(Gupta et al. 2006) Gupta, P., et al.: Gate-length biasing for runtime-leakage control. IEEE Transactions on Computer-Aided Design of Integrated Circuits and Systems 25(8), 1475–1485 (2006)

(Hailin et al. 2005) Jiang, H., et al.: Benefits and costs of power-gating technique. In: Proceedings of 2005 IEEE International Conference on Computer Design: VLSI in Computers and Processors, ICCD 2005, October 2-5, pp. 559–566 (2005)

(Hampel et al. 2009) Hampel, S.K., et al.: Inductorless 1-10.5 GHz Wideband LNA for Multistandard Applications. In: IEEE Asian Solid-State Circuits Conference (November 2009)

(Harrison and Charles 2003) Harrison, R.R., Charles, C.: A Low-Power Low-Noise CMOS Amplifier for Neural Recording Applications. IEEE Journal of Solid-State Circuits 38(6), 958–965 (2003)

(Ho et al. 2003) Ho, C.C., et al.: A 2.4 GHz low phase noise VCO fabricated by 0.18μm pMOS technologies. In: Proc. of IEEE Int. Symp. VLSI Tech., p. 144 (2003)

(Hobbs and Williams 2008) Hobbs, J.S., Williams, T.W.: Reaching the limits of low power design. In: Asia and South Pacific Design Automation Conference, ASPDAC 2008, March 21-24, pp. 732–735 (2008)

(Hsieh et al. 2010) Hsieh, W.-C., Hwang, W.: Adaptive Power Control Technique on Power-Gated Circuitries. IEEE Transactions on Very Large Scale Integration (VLSI) Systems 99, 1–14 (2010)

(Hsieh et al. 2011) Hsieh, H.-H., Chen, H.-S., Hung, P.-H., Lu, L.-H.: Experimental 5-GHz RF Frontends for Ultra-Low-Voltage and Ultra-Low-Power Operations. IEEE Transactions on Very Large Scale Integration (VLSI) Systems 19(4), 705–709 (2011)

(Huaizhi and Wong 2007) Huaizhi, W., Wong, M.D.F.: Improving Voltage Assignment by Outlier Detection and Incremental Placement. In: 44th ACM/IEEE Design Automation Conference, DAC 2007, June 4-8, pp. 459–464 (2007)

(Huaizhi et al. 2006) Huaizhi, W., et al.: Timing-constrained and voltage-island-aware voltage assignment. In: 43rd ACM/IEEE Design Automation Conference, pp. 429–432 (2006)

(Huaizhi and Wong 2009) Huaizhi, W., Wong, M.D.F.: Incremental Improvement of Voltage Assignment. IEEE Transactions on Computer-Aided Design of Integrated Circuits and Systems 28(2), 217–230 (2009)

(Ishida et al. 2005) Ishida, K., et al.: Subthreshold-leakage suppressed switched capacitor circuit based on super cut-off CMOS (SCCMOS). In: IEEE International Symposium on Circuits and Systems, ISCAS 2005, May 23-26, vol. 4, pp. 3119–3122 (2005)

(Kawaguchi et al. 2000) Kawaguchi, H., et al.: A Super Cut-Off CMOS (SCCMOS) Scheme for 0.5-V Supply Voltage with Picoampere Stand-By Current. IEEE Journal of Solid-State Circuits 35(10), 1498–1500 (2000)

(Karimi and Sarrafzadeh 2010) Karimi, Z., Sarrafzadeh, M.: Fine-grained post placement voltage assignment considering level shifter overhead. In: 2010 18th IEEE/IFIP VLSI System on Chip Conference (VLSI-SoC), September 27-29, pp. 73–78 (2010)

(Khalek et al. 2007) Khalek, F., Yusoff, Z., Sulaiman, M.-S.: Low Power Techniques for a Mixed-Signal Circuit. In: International Symposium on Integrated Circuits, ISIC 2007, September 26-28, pp. 73–76 (2007)

(Khan et al. 2006) Khan, A., et al.: A 90-nm Power Optimization Methodology With Application to the ARM 1136JF-S Microprocessor. IEEE Journal of Solid-State Circuits 41(8), 1707–1717 (2006)

(Karanikolas 1996) Karanicolas, A.C.: A 2.7 V 900 MHz CMOS LNA and mixer. IEEE Journal of Solid State Circuits 31, 50 (1996)

(Kulkarni and Sylvester 2003) Kulkarni, S.H., Sylvester, D.: Fast and energy-efficient asynchronous level converters for multi-VDD design [CMOS ICs]. In: Proceedings of the IEEE International SOC Conference [Systems-on-Chip], September 17-20, pp. 169–172 (2003)

(Lee et al. 2006) Lee, J., Potts, J., Spears, E.: DC/DC converter controlled power amplifier module for WCDMA applications. In: 2006 IEEE Radio Frequency Integrated Circuits (RFIC) Symposium, June 11-13, p. 4 (2006)

(Lee and Choi 2000) Lee, S.G., Choi, J.K.: Current-Reuse Bleeding Mixer. Electronic Letters 36(8), 696 (2000)

(Lee 1998) Lee, T.H.: The Design of CMOS Radio Frequency Integrated Circuits. Cambridge University Press, Cambridge (1998)

(Lee et al. 2009) Lee, M.-C., et al.: An efficient wakeup scheduling considering resource constraint for sensor-based power gating designs. In: IEEE/ACM International Conference on Computer-Aided Design - Digest of Technical Papers, ICCAD 2009, November 2-5, pp. 457–460 (2009)

(Li et al. 2010) Li, M., et al.: A single-chip 2.4GHz double cascode power amplifier with switched programmable feedback biasing under multiple supply voltages in 65nm CMOS for WLAN application. In: 2010 IEEE Radio Frequency Integrated Circuits Symposium (RFIC), May 23-25, pp. 391–394 (2010)

(Liscidini et al. 2006) Liscidini, A., et al.: Single-Stage Low-Power Quadrature RF Receiver Front-End: The LMV Cell. IEEE Journal of Solid State Circuits 41(12), 2832 (2006)

(Makino et al. 1998) Makino, H., et al.: An auto-backgate-controlled MT-CMOS circuit. In: 1998 Symposium on VLSI Circuits, Digest of Technical Papers, June 11-13, pp. 42–43 (1998)

(Meng et al. 2007) Meng, L.K., et al.: A 2.4GHz Ultra Low Power Subthreshold CMOS Low-Noise Amplifier. Microwave and Optical Technology Letters 49, 743 (2007)

(Mutoh et al. 1995) Mutoh, S., et al.: 1-V power supply high-speed digital circuit technology with multithreshold-voltage CMOS. IEEE Journal of Solid-State Circuits 30(8), 847–854 (1995)

(Okano et al. 2006) Okano, H., et al.: Supply Voltage Adjustment Technique for Low Power Consumption and Its Application to SOCs with Multiple Threshold Voltage CMOS. In: 2006 Symposium on VLSI Circuits, Digest of Technical Papers, pp. 208–209 (2006)

(Panic et al. 2008) Panic, G., et al.: Architecture of a Power-Gated Wireless Sensor Node. In: 11th EUROMICRO Conference on Digital System Design Architectures, Methods and Tools, DSD 2008, September 3-5, pp. 844–849 (2008)

(Panic et al. 2008 (2)) Panic, G., et al.: Power gating in wireless sensor networks. In: 3rd International Symposium on Wireless Pervasive Computing, ISWPC 2008, May 7-9, pp. 499–503 (2008)

(Pasha et al. 2010) Pasha, M.A., et al.: A novel approach for ultra low-power WSN node generation. In: Signals and Systems Conference (ISSC 2010), IET Irish, June 23-24, pp. 204–209 (2010)

(Quemada et al. 2009) Quemada, C., et al.: Design Methodology for RF CMOS Phase Locked Loops. Artech House, Boston (2009)

(Ping et al. 2010) Huang, P., et al.: A brief survey on power gating design. In: 2010 10th IEEE International Conference on Solid-State and Integrated Circuit Technology (ICSICT), November 1-4, pp. 788–790 (2010)

(Razavi 2001) Razavi, B.: "Design of Analog CMOS Integrated Circuits. McGraw-Hill, New York (2001)

(Rippke et al. 2005) Rippke, I.A., et al.: A single-chip variable supply voltage power amplifier. In: 2005 IEEE Radio Frequency integrated Circuits (RFIC) Symposium, Digest of Papers, June 12-14, pp. 255–258 (2005)

(Ryu et al. 2007) Ryu, S., et al.: A 10-bit 50-MS/s Pipelined ADC With Opamp Current Reuse. IEEE Journal of Solid State Circuits 42(3), 475 (2007)

(Sedra and Smith 2004) Sedra, A.S., Smith, K.C.: Microelectronic Circuits. Oxford University Press, Oxford (2004)

(Sengupta and Saleh 2007) Sengupta, D., Saleh, R.: Constraint-based voltage island partitioning. In: IEEE Northeast Workshop on Circuits and Systems, NEWCAS 2007, August 5-8, pp. 1050–1053 (2007)

(Shibata et al. 2010) Shibata, N., et al.: A High-Speed Low-Power Multi-VDD CMOS/SIMOX SRAM With LV-TTL Level Input/Output Pins—Write/Read Assist Techniques for 1-V Operated Memory Cells. IEEE Journal of Solid-State Circuits 45(9), 1856–1869 (2010)

(Sirisantana et al. 2000) Sirisantana, N., et al.: High-performance low-power CMOS circuits using multiple channel length and multiple oxide thickness. In: Proceedings of the 2000 International Conference on Computer Design, pp. 227–232 (2000)

(Sirisantana et al. 2004) Sirisantana, N., et al.: Low-power design using multiple channel lengths and oxide thicknesses. IEEE Design & Test of Computers 21(1), 56–63 (2004)

(Sjalander et al. 2005) Sjalander, M., et al.: A low-leakage twin-precision multiplier using reconfigurable power gating. In: IEEE International Symposium on Circuits and Systems, ISCAS 2005, May 23-26, vol. 2, pp. 1654–1657 (2005)

(Sjoland et al. 2003) Sjöland, H., et al.: A Merged CMOS LNA and Mixer for a WCDMA Receiver. IEEE Journal of Solid State Circuits 38(6), 1045 (2003)

(Sylvester and Kaul 2001) Sylvester, D., Kaul, H.: Future performance challenges in nanometer design. In: Proceedings of Design Automation Conference, pp. 3–8 (2001)

(Taris et al. 2009) Taris, T., et al.: Current reuse topology in UWB CMOS LNA. Analog Integrated Circuits and Signal Processing, 149 (online, 2009)

(Tawfik and Kursun 2007) Tawfik, S.A., Kursun, V.: Multi-Vth Level Conversion Circuits for Multi-VDD Systems. In: IEEE International Symposium on Circuits and Systems, ISCAS 2007, May 27-30, pp. 1397–1400 (2007)

(Tsividis 1998) Tsividis, Y.: Operation and Modeling of the MOS Transistor, 2nd edn. McGraw-Hill, Boston (1998)

(Valentian and Beigne 2008) Valentian, A., Beigne, E.: Automatic Gate Biasing of an SCCMOS Power Switch Achieving Maximum Leakage Reduction and Lowering Leakage Current Variability. IEEE Journal of Solid-State Circuits 43(7) (July 2008)

(Vidojkovic et al. 2005) Vidojkovic, V., et al.: A Low-Voltage Folded-Switching Mixer in 0.18-μm CMOS. IEEE Journal of Solid State Circuits 40(6) (June 2005)

(Vittoz and Feelrath 1997) Vittoz, E., Fellrath, J.: CMOS analog integrated circuits based on weak inversion operations. IEEE Journal of Solid-State Circuits 12(3), 224–231 (1977)

(Vittoz 2003) Vittoz, A.: Weak inversion in analog and digital circuits. In: CCCD Workshop (2003)

(Walling et al. 2009) Walling, J.S., et al.: Multiple supply (class-G) linear modulator and PA for non-CE modulation. In: Radio Frequency Integrated Circuits Symposium, RFIC 2009, June 7-9, pp. 587–590. IEEE, Los Alamitos (2009)

(Wang et al. 2006) Wang, J., et al.: Design of STR level converters for SoCs using the multi-island dual-VDD design technique. In: Proceedings of 2006 IEEE International Symposium on Circuits and Systems, ISCAS 2006 (2006)

(Wong 2007) Wong, M.D.F.: Low power design with multi-Vdd and voltage islands (Abstract). In: 7th International Conference on ASIC, ASICON 2007, October 22-25, pp. 13–25 (2007)

(Zhihua et al. 2008) Wang, Z., et al.: An energy-efficient ASIC with real-time work-on-demand for wireless body sensor network. In: IEEE International Conference on Electron Devices and Solid-State Circuits, EDSSC 2008, December 8-10, pp. 1–6 (2008)

6 RF Amplifier Design

This chapter deals with the design of the basic amplifying stages. It also provides the introductory tools required for the design of any kind of amplifier (VGA, LNA, BUFFERS, etc). Due to its importance for CMOS ICs, the low noise amplifier is deeply studiedi this chapter.

Section 6.1 presents the basic concepts of this circuit and its basic configurations. Section 6.2 deals with the different topologies and the impact on power consumption. Firstly the difference between single ended and differential architectures is presented, and secondly different architectural configuration hints are given regarding to the load and input matching considerations respectively. Section 6.3 integrates more design constraints usually found during the design process that trade-off the low-power performance, such as gain, linearity, noise figure, bandwidth and stability and finally, Section 6.4 extensively presents some examples, including design tips, photographs of layouts and characterization results.

6.1 Basic Stages Fundamentals

The MOSFET are three-terminal devices (4 if the substrate is also considered). These three terminals are generally connected so as the transistor itself is a 2-port network. Therefore, three different single-transistor amplifier configurations can be formed, depending on which of the three transistor terminals is used as signal ground. Such basic configurations are called common source, common drain (source follower), and common gate, deppending which of the terminals is common to noth input/output ports.

The input and output resistance characteristics of amplifiers are important in determining loading effects. These parameters, as well as voltage and current gains, are the most representative in terms of amplifier performance. Such effects, as well as noise and linearity performance have been deeply analyzed in the literatura (refs) by using the small signal models and different circuit approaches. However, they will be here very basically introduced in order to get an insigth about their main characteristics for the implementation of different amplifier (e.g. LNA) topologies, at will be described in section 6.2.

This section starts with some basic expressions about NMOS transistors performance for a better understanding of the main transistor configurations described thereafter. Then, each configuration is briefly summarized within following subsections, and finaly their main parameters are qualitatively compared.

U. Alvarado et al.: Low Power RF Circuit Design in CMOS Tech., LNEE 104, pp. 87–127.
springerlink.com

6.1.1 NMOS Transistor Basic Expressions

In this point we present a brief outline of basic expressions and figures that are used to perform simple analysis of the different circuits. As an starting point, Fig. 6.1 shows the symbol and terminal names that will be used throughout this chapter for NMOS and PMOS transistors.

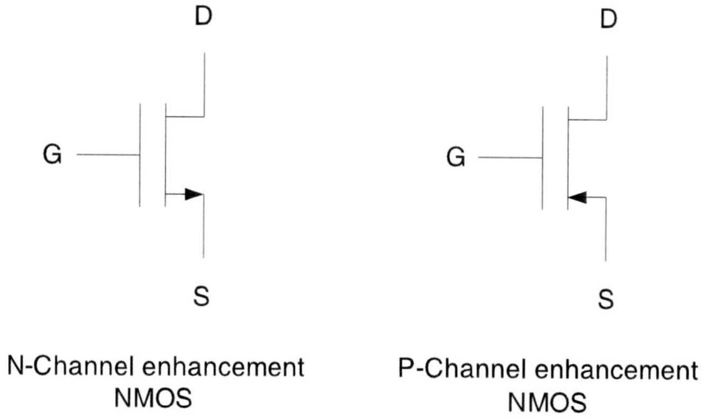

Fig. 6.1 MOS transistor symbols

Regarding the I_D vs. V_{GS} behavior, Fig. 6.2 shows the transition between subthreshold and saturation operating zones. Due to the crucial importance of subthreshold operation in low power application, this topic is covered with greater detail in subsequent paragraphs (see chapter 5).

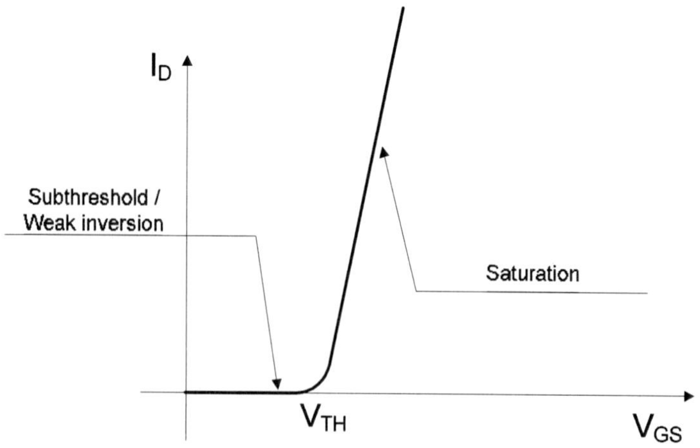

Fig. 6.2 $I_D$vs V_{GS} of a typical NMOS transistor

The last of the basic curves (see Fig. 6.3) shows the different alternatives across the V_{DS} variation range. We will consider two options: triode and saturation.

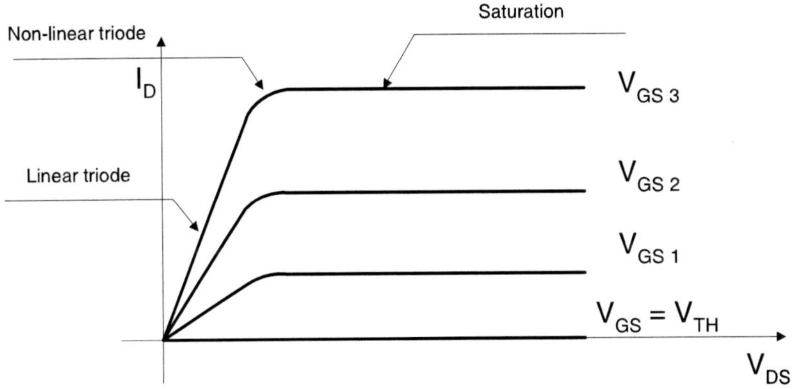

Fig. 6.3 $I_D$vs V_{DS} of a typical NMOS transistor

Simple DC analysis may be carried out using the equations displayed above.

$$I_{D\ TRIODE} = \mu_n C_{ox} \frac{W}{L}\left[(V_{GS} - V_{TH})V_{DS} - \frac{1}{2}V_{DS}^2\right] \quad \text{Eq. 6.1}$$

$$I_{D\ SATURATION} = \frac{1}{2}\mu_n C_{ox} \frac{W}{L}(V_{GS} - V_{TH})^2 \quad \text{Eq. 6.2}$$

Where μ_n is theelectron mobility, C_{ox} is thecapacity per unit gate area, W and L are the channel width and length respectively, V_{GS} is the gate-source voltage and V_{TH} is the threshold voltage. From Eq. 6.2, an expression of the transconductance can be derived:

$$g_m = \mu_n C_{ox} \frac{W}{L}(V_{GS} - V_{TH}) \quad \text{Eq. 6.3}$$

For AC small signal fast analysis, the high frequency simplified model shown in Fig. 6.4 is recommended:

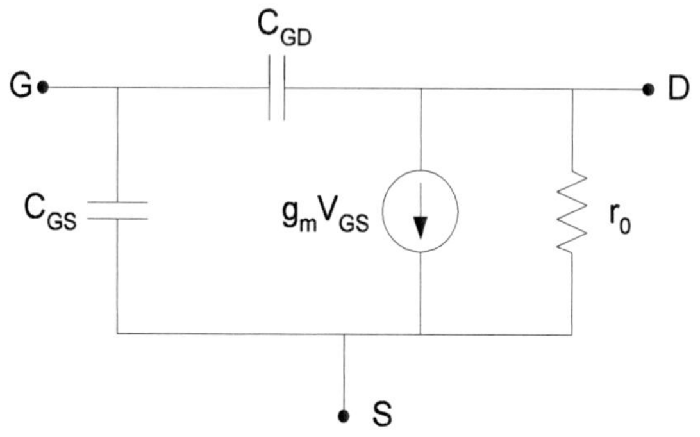

Fig. 6.4 Simplified high frequency model of NMOS transistor

The values of C_{GS} and C_{GD} can be estimated with the expressions shown in Table 6.1.

Table 6.1 Estimation of capacitances of the high frequency model

	Cutoff	Saturation	Triode
C_{gs}	0	$\frac{2}{3}WLC_{ox}$	$\frac{1}{2}WLC_{ox}$
C_{gd}	0	0	$\frac{1}{2}WLC_{ox}$

A very useful expression for Transition Frequency estimation can be obtained from a straightforward inspection of the simplified HF model:

$$f_T = \frac{g_m}{C_{gs} + C_{gd}} \qquad \text{Eq. 6.4}$$

6.1.2 Common Source Configuration

This stage is present in almost every amplification chain, due to its good balance among the different characteristics (input/output impedances, gain, noise, etc.). Fig. 6.5 shows the basic schematic circuit, without DC polarization components. R_S and R_L are the equivalent series resistance and load respectively.

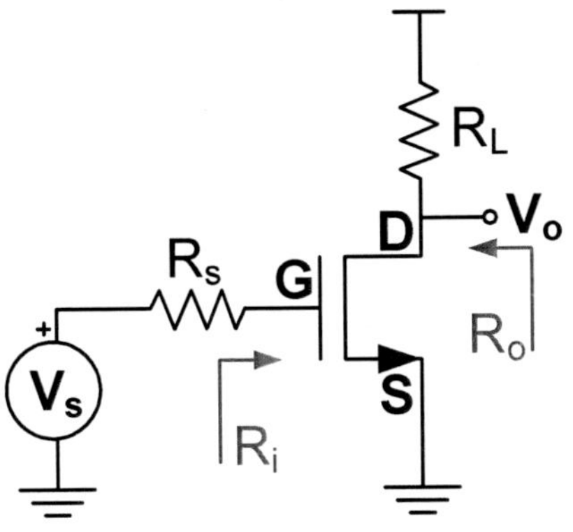

Fig. 6.5 Common source basic schematic circuit

In order to derive the different expressions that characterise the performance of the common-source configuration, the most simple small-signal transistor model is used in the schematic of Fig. 6.5, as shown in Fig. 6.6.

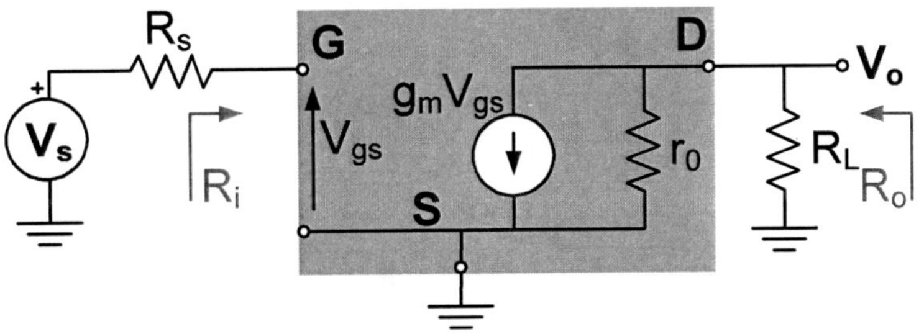

Fig. 6.6 Common source configuration. Small-signal model

The following set of equations show the performance of this configuration, in terms of voltage gain, input impedance and output impedance.

$$A_V = \frac{V_o}{V_s} = -g_m \left(R_L \,\|\, r_0 \right) \qquad \text{Eq. 6.5}$$

$$R_i = \infty \qquad \text{Eq. 6.6}$$

$$R_o = r_0 \parallel R_L \qquad \text{Eq. 6.7}$$

One of the main drawbacks of this basic cell is the relatively poor high frequency behavior, due to the Miller effect. In fact, parasitic capacitance C_{gd}(see Fig. 6.4) is connected between two nodes of a high voltage gain network. Hence, the high frequency dominant pole is usually due to this effect, and can be estimated through the next equation.

$$f_{3dB} = \frac{1}{2\pi R_S} \frac{1}{C_{GS} + C_{GD}\left(1 + g_m R_L\right)} \qquad \text{Eq. 6.8}$$

In order to overcome this problem, there are some alternative schemes that might be used, as it will bedescribed in section 6.2.

6.1.3 Common Drain Configuration

Fig. 6.7 depicts the common drain basic cell, also known as source follower. Once again, DC bias is omitted, and R_S, R_L include all the resistive effects associated to the input and output ports.

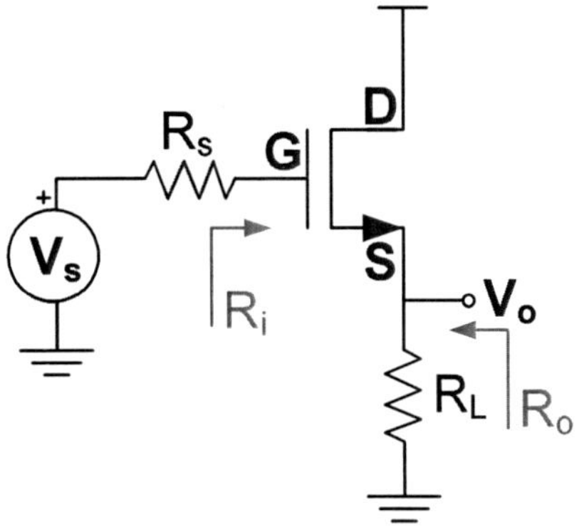

Fig. 6.7 Common drain basic schematic circuit

In order to derive the different expressions that characterise the performance of the common-drain configuration, the most simple small-signal transistor model is used in the schematic of Fig. 6.7, as shown in Fig. 6.8.

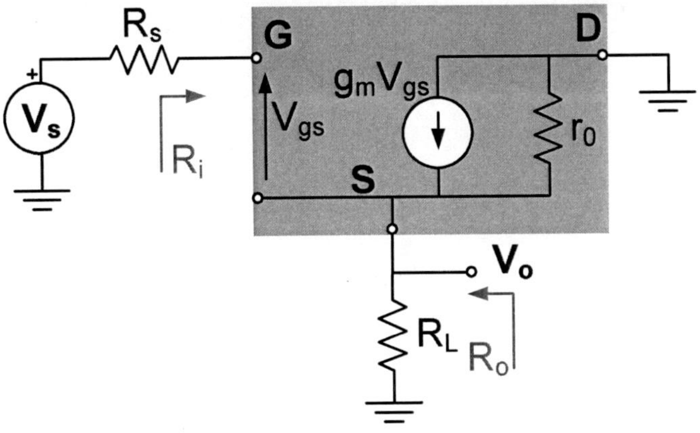

Fig. 6.8 Common-drain configuration. Small-signal model

The following set of equations represent the performance of the common-drain configuration in terms of voltage gain, and input / output impedances respectively. On the other hand, Eq. 6.12 accounts for the dominant frequency pole (3dB bandwidth):

$$A_v = \frac{V_o}{V_s} = \frac{g_m r_0}{1 + g_m r_0 + \frac{r_0}{R_L}} \qquad \text{Eq. 6.9}$$

$$R_i = \infty \qquad \text{Eq. 6.10}$$

$$R_o = \frac{1}{g_m + \frac{1}{r_0} + \frac{1}{R_L}} \qquad \text{Eq. 6.11}$$

$$f_{3dB} = \frac{1}{2\pi}\left[\left(C_{GS} + \frac{C_{GD}}{1 + g_m R_L}\right) R_S\right]^{-1} \qquad \text{Eq. 6.12}$$

As it is apparent, no voltage gain is provided, but the upper cut-off frequency is sensibly higher than the common source cell. The output impedance is also lower, and therefore this is the scheme usually selected to implement the output stage of an amplification chain.

6.1.4 Common Gate Configuration

A basic common gate cell is shown in Fig. 6.9.

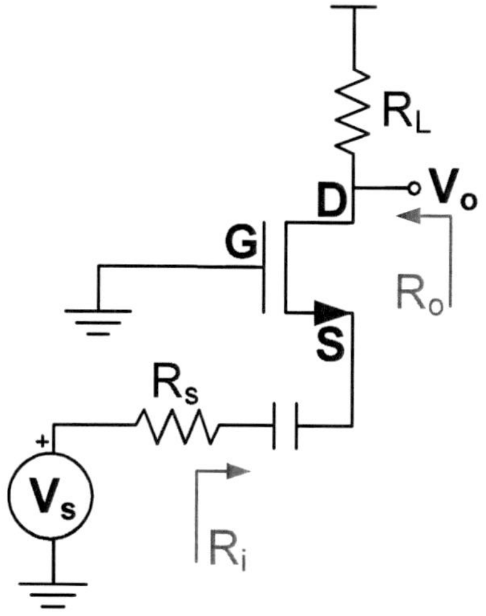

Fig. 6.9 Common gate basic schematic circuit

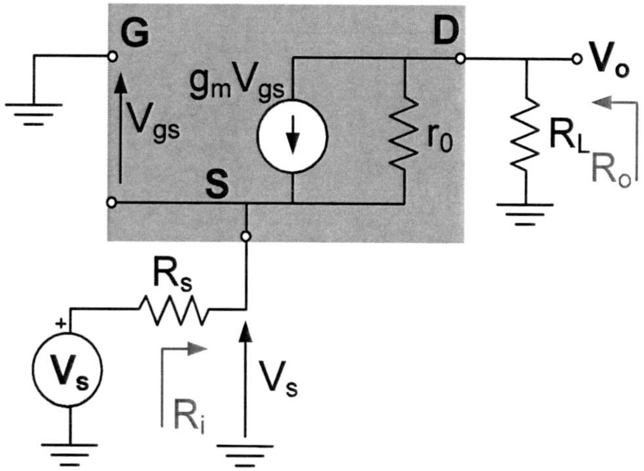

Fig. 6.10 Common-gate configuration. Small-signal model

A small signal analysis with the simplified model of Fig. 6.10 is used once again to derive the following set of equations, describing the voltage gain, input impedance and output impedance respectively. In this case the frequency poles are determined by Eq. 6.16:

$$A_v = \frac{V_o}{V_s} = R_L \frac{1 + g_m r_0}{R_L + R_S + r_0 (1 + g_m R_S)} \qquad \text{Eq. 6.13}$$

$$R_i = \frac{r_0 + R_L}{1 + g_m r_0} \qquad \text{Eq. 6.14}$$

$$R_o = R_L \, || \left[R_S (1 + g_m r_0) + r_0 \right] \qquad \text{Eq. 6.15}$$

$$f_{p1} = \frac{1}{2\pi} \frac{1}{C_{gs} \left(\dfrac{1}{g_m} \, || \, R_S \right)} \quad ; \quad f_{p2} = \frac{1}{C_{gd} R_L} \qquad \text{Eq. 6.16}$$

Where f_{p1} and f_{p2} are the high frequency poles. In this case, the parasitic capacitance feedback is avoided; hence the bandwidth is the highest of the three basic cells. However, as the input signal is connected to the source terminal and the gate is grounded, this circuit loses the high input impedance usually associated to MOS transistors.

6.1.5 Comparison of the Basic Configurations

Table 6.2 is a summary the main characteristics of the three amplifier configurations, in terms of their voltage and current gains, and input/output impedances respectively.

Table 6.2 Transistor configurations summary

	Voltagegain	Currentgain	Input impedance	Output impedance
CommonSource	High	∞	∞	High
CommonDrain	R_S	≈ 1	Low	R_L
CommonGate	≈ 1	∞	∞	Low

The specific characteristics of these single-stage amplifiers are used in the design of multistage amplifiers.

6.2 Amplifier Topologies

In this subsection we present the most representative topologies used in the design of amplifiers for high frequency. The objective of this section is not the in deepth discussion of any of these topologies, as further information can be found in the literature related to the basic electronic circuits books included in the references (Caverly 2007), (Horenstein 1995), (Lee 1998), (Sedra and Smith 2009).

6.2.1 Cascoded Amplifier

This technique is a frequently used way to reduce the Miller effect, especially in common source cells. The cascode transistor (G2) lowers the voltage gain of G1 to a value close to one. Therefore, it reduces the Miller effect. On the other hand, G2 transfers (ideally) the current I_{D1} to R_L, with an overall gain similar to the common source basic cell.

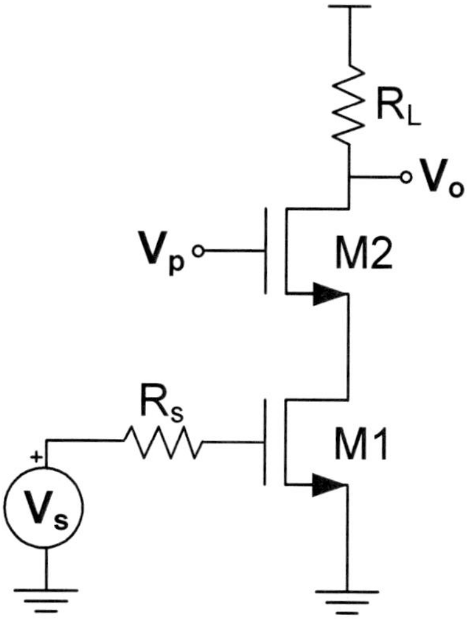

Fig. 6.11 Basic cascode cell (DC bias not shown)

As a consequence, high frequency dominant pole is independent of M1 transconductance, thus independent also of the gain.

$$f_p = \frac{1}{R_S\left(C_{gs}+2C_{gd}\right)} \qquad \text{Eq. 6.17}$$

Ideally, the high frequency behavior of this stage can be determined just by the optimization of the transistor acting as trasconductor (M1). However, second-order effects may change this situation.

6.2.2 Tuned Load: LC-Tank

One of the most used topologies within RF front-ends is the tuned amplifier. In this scheme, the resistive load is substituted by an LC-tank. There are many reasons that indicate its frequent use, among them:

- Low noise and good linearity of the LC-tank compared to the resistive or active load
- High gain in the frequency of interest
- High rejection of out-of-band frequencies

The basic diagram is shown in Fig. 6.12.

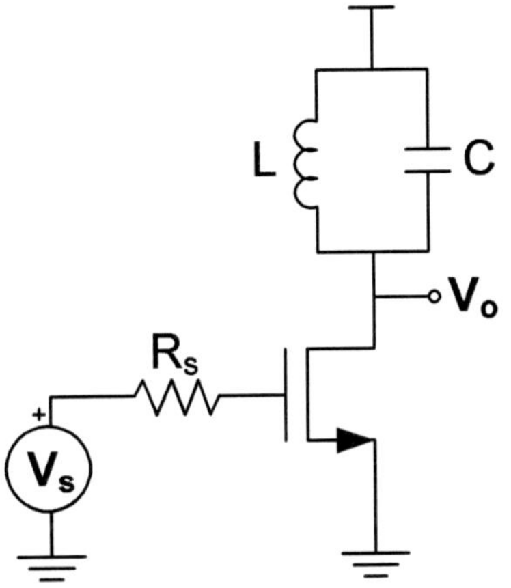

Fig. 6.12 Tuned amplifier with LC-tank

The frequency of the peak can be easily determined by the calculation of the resonance point of the LC-tank.

$$f_o = \frac{1}{2\pi}\frac{1}{\sqrt{LC}} \quad \text{Eq. 6.18}$$

This is a good approximation for discrete components; however, special care must be taken when we deal with integrated circuits for high frequencies. Regarding the capacitor, currently available technologies provide integrated capacitors with relatively small values, especially when we try to minimize the area of the circuit. Due to this fact, the parasitic capacitances of metal tracks or other components add a significant amount of its effect to the implemented capacitor, lowering the expected resonance.

But the situation is far more complex with integrated inductors. The different metal tracks and layers give rise to a great number of undesired parasitic effects. Fig. 6.13 shows one of the most commonly used models for integrated inductors analysis: The π-model.

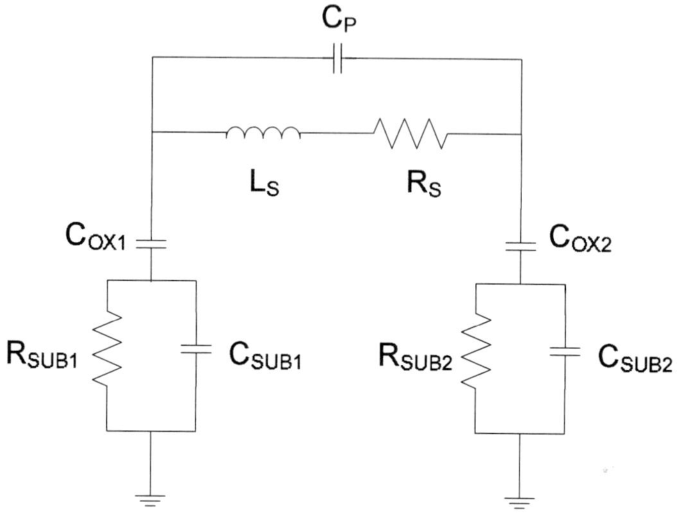

Fig. 6.13 Integrated inductor π-model

In this model:

- C_p: Capacitance between the metal tracks and the interior connection to the inductor
- L_S: Main inductance of the inductor, including the self-inductance plus the mutual inductance between tracks of a same level and between tracks of different layers
- R_S: Series resistance of the inductor, including ohmic losses of the metal tracks, particular effects and losses due to currents induced in the metal
- $C_{ox1}$and C_{ox2}: Capacitance between the spiral metal track and the substrate.
- $R_{sub1}$and R_{sub2}:Ohmic losses from the substrate produced by the induced currents
- $C_{sub1}$and C_{sub2}: Capacitive effects of the substrate due to its semiconductor character.

Due to its simplicity, this model is used to perform simple first order calculations. But we should take into account that it is a narrow band model, only reliable in a frequency range of some hundreds of MHz around its centre frequency (Aguilera and Berenguer, 2003).

6.2.3 *Active Load*

This case takes advantage of the high dynamic impedance of the MOS current source, obtaining a good compromise between voltage gain and occupied silicon area.

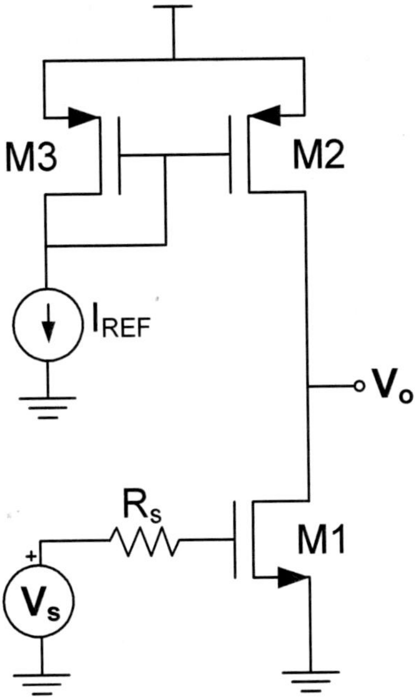

Fig. 6.14 Active loaded common source basic cell

The main drawback of this cell is the relatively poor high frequency behavior, due to the increase in total capacitance seen at Q_1's output produced by Q_2 and Q_3 transistors. One solution for this problem is the introduction of a cascode transistor. With this configuration, the dominant pole is caused mainly by the cascode transistor and the total capacitance at its output, as it is shown in the next expression:

$$f_p = \frac{1}{2\pi}\frac{1}{\left(C_{gsC} + C_L\right)R_L} \qquad \text{Eq. 6.19}$$

Where C_{gsC} is referred to the cascade device. In any case, this cell is not used in high frequency applications, but mainly in baseband circuits.

6.2.4 Negative Feedback Estructures

Negative feedback is a technique widely used in electronic design, and especially in amplification chains. There are many benefits that can be obtained with an appropriate introduction of these techniques, including:

- Bandwidth extension
- Impedance matching
- Noise cancellation

- Linearity improvement
- Unconditional stability

Fig. 6.15 depicts the general feedback structure. In this figure block A represents the amplifier, while β is the feedback network.

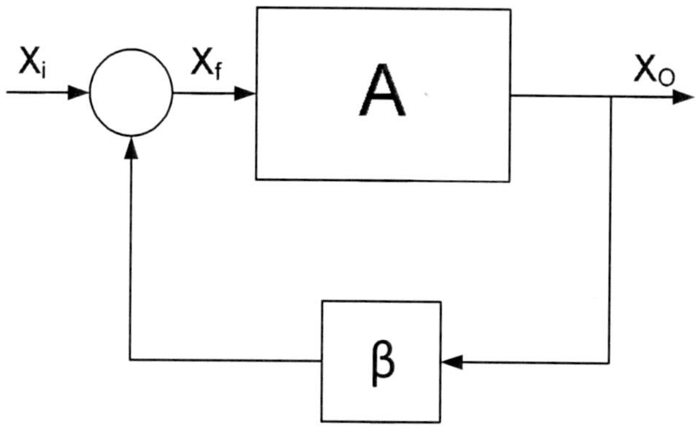

Fig. 6.15 General feedback structure

The feedback ratio is defined as the ratio between the feedback signal and the output signal:

$$\beta = \frac{X_f}{X_o} \quad \text{Eq. 6.20}$$

Feedback signal is subtracted from the input, i.e.:

$$X_f = X_i - \beta X_o \quad \text{Eq. 6.21}$$

Therefore, the gain of the complete system can be written as follows:

$$A_f = \frac{A}{1 + A\beta} \quad \text{Eq. 6.22}$$

The practical implementation of this technique depends mainly on the type of signal of interest. There are two options: current and voltage, giving four different topologies, as indicated in the next table.

Although all these four alternatives can be found in RF circuits, there are two particular cases more frequently used than the rest: shunt feedback (shunt-series) and resistive/inductive degeneration (series-series). In the next sub-subsections we briefly discussed both.

Table 6.3 Four Feedback Amplifier Topologies

X_i	X_O	Amplifiertype
Voltage	Voltage	Series-Shunt
Current	Current	Shunt-Series
Voltage	Current	Series-Series
Current	Voltage	Shunt-Shunt

6.2.4.1 Resistive/Inductive Degeneration

The introduction of a degeneration resistor is a straightforward method to increase the input impedance and improve linearity, but at the cost of a higher noise figure and a lower gain.

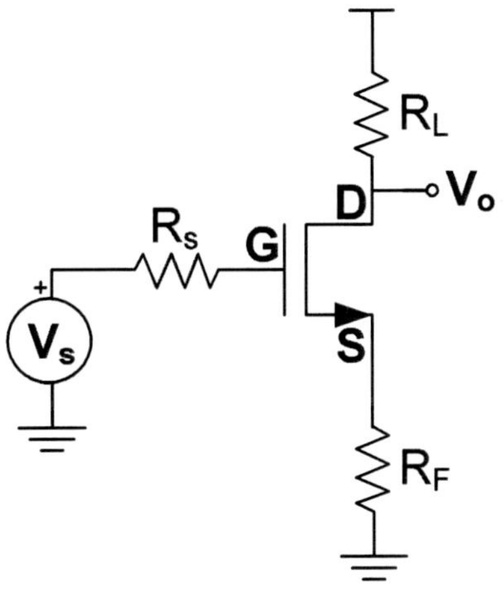

Fig. 6.16 Resistive degeneration of a common source stage

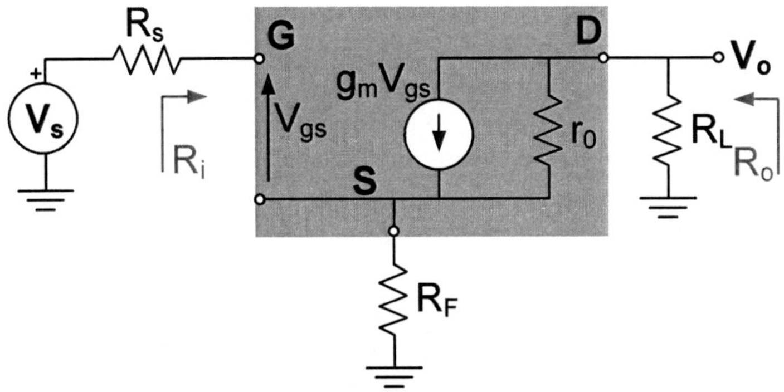

Fig. 6.17 Resistive degeneration of a common source stage. Small signal model

The following set of equations show the voltage gain and input/output impedances of a source-deferenated common-source configuration with the small signal model of Fig. 6.17.

$$A_v = \frac{V_o}{V_s} = \frac{-R_L g_m}{1 + g_m R_F} \approx \frac{-R_L}{R_F} \qquad r_0 >> \qquad \text{Eq. 6.23}$$

$$R_i = \infty \qquad \text{Eq. 6.24}$$

$$R_o = R_L \frac{R_F + r_0 \left(1 + g_m R_F\right)}{R_L + R_F + r_0 \left(1 + g_m R_F\right)} \qquad \text{Eq. 6.25}$$

In RF applications, the increase in the noise contribution is quite often unacceptable, and inductive degeneration is preferred. Fig. 6.18 pictures an example.

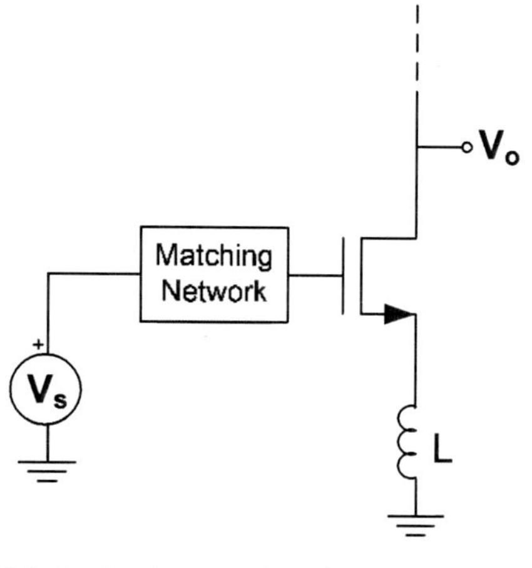

Fig. 6.18 Inductive degeneration of a common source stage

With the introduction of the inductance, the phase difference between input voltage and current could propose an additional problem. However, it is easy to show that, if an appropriate matching network is also included, a total control over the input impedance is achieved (Lee 1998).

6.2.4.2 Shunt Feedback

This technique is included in the design of broadband amplifiers, as it combines an extended bandwidth with stability in a desired frequency range.

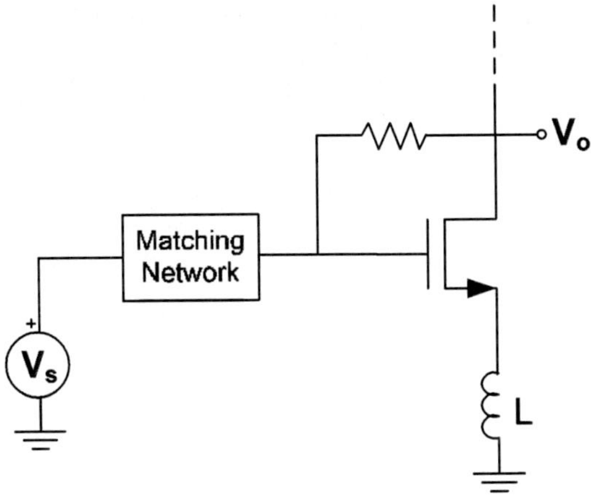

Fig. 6.19 Shunt feedback network applied to a common source stage

Once again, the main drawback of this topology is the noise contribution of the feedback network. The situation may be improved with the introduction of an inductive degeneration, but assuming an important reduction in the frequency range.

6.3 LNA Low Power Design Considerations

This section integrates the LNA design constraints usually found during the design process but focused on the low-power performance. The parameters studied for the design of this block are gain, linearity, noise figure, impedance matching and stability; but these are studied through the 6 tools presented in that section. Nevertheless, first of all, the basic stage for the core of the low noise amplifier shall be selected.

As introduced in the first section of this chapter, two topologies could be taken into consideration for the core of an LNA: the common source (CS) and the common gate (CG) amplifiers. With their basic configuration, independently of their load, both Eq. 6.26 and Eq. 6.27 show the simplified expressions for the current needed for a 3 dB Noise Figure (Abidi et al. 2000).

$$I_{DRAIN\_CommonSource} = \frac{g_m V_{eff}}{2} \approx \frac{V_{eff}}{10RGS_{CS}} \approx \frac{V_{eff}}{10Z_{INCS}} \qquad \text{Eq. 6.26}$$

$$I_{DRAIN\_CommonGate} = \frac{g_m V_{effCG}}{2} \approx \frac{V_{effCG}}{2Z_{INCG}} \approx \frac{V_{eff}}{2\sqrt{Z_{INCS} Z_{INCG}}} \qquad \text{Eq. 6.27}$$

Where V_{eff} and V_{effCG} are the overdrive voltages of the CS and CG topologies, respectively. RGS_{CS} is the inversion layer resistance of the CS scheme, Z_{INCS} is the input impedance of the CS amplifier (typically 50 Ω for RF designs) and Z_{INCG} is the input impedance of the CG amplifier.

At first sight, the CS topology seemed to improve the CG amplifier architecture in the current consumption by a factor of 5. Nevertheless, the CG had the possibility to integrate an impedance transformer at its input. Abidi, et al., deduced that with the Z-transformer the resulting current diminished by a factor of the squared root of the Z_{INCG}/Z_{INCS} ratio. A minimum of 25 is needed to overcome the five times topological disadvantage. The simplest of these kinds of transformers is a LC narrow band circuit, and hence, would require a 25 times ratio on the Q of the inductors of this transformer. Consequently, at the band of interest and for integrated designs, the high frequency and the parasitics advised the common source configuration for the core of the LNA.

From this starting point, the current consumption design optimization for this block has several points to consider in open-loop small-signal circuits such as the LNA. Six LNA generic design issues are here presented.

6.3.1 Inductive Degeneration

The spurious free Dynamic Range (DR) of an LNA is limited at the upper end by the largest signal that is undistorted at the output of the device. The lower-end boundary is defined by the noise spectral density, integrated across the band of interest. The difference between these two power levels is the DR of the device, its available working region. Any widening of this interval has a positive repercussion on the linearity of the LNA.

A theoretical method to extend the DR is the inductive degeneration, decoupling the DR from the power consumption, which means a higher IIP3 with the same current consumption. Unfortunately, the addition of these inductors affects the frequency response of the LNA and the gain in comparison to the basic stage, but electronics requires trade-offs.

6.3.2 Q - Passive Devices

The quality factor (Q) of the passive devices of the circuit (if any) has a great influence in the Performance Vs. Power dissipation ratio. As explained in the second section of this chapter, one of the most used basic topologies for LNAs uses an LC tank as a load. As said, the load inductor resonates with the capacitance in the frequency band of interest. The resulting peak voltage gain depends on the impedance of this tank, and thisZ_{tank} is limited by the Q of the tank, which is mainly fixed by the Q of the inductor. The next equation reveals the relation between the LNA's gain and the Q.

$$Gain_{LNA} = g_m Z_{\tan k} = g_m Q\omega L = g_m \frac{Q}{\omega C} = \frac{2QI}{\omega C V_{eff}} \qquad \text{Eq. 6.28}$$

Where ω is the angular frequency ($\omega = 2\pi f$), C and L are the values of the capacitance and the inductance of the tank respectively, I is the drain current of the LNA and V_{eff} is the overdrive voltage applied at the gate of the amplifying transistor. This equation reflects that to lower the power dissipation for a given gain, the quality factor of the passive elements has to be increased.

6.3.3 *Transistor Polarization*

Another outcome from Eq. 6.28 (la anterior) is the dependence of the gain on V_{eff}. In fact, the lowering of the gate DC biasing voltage increased the Gain/Current ratio (see chapter 5). Moreover, the decreasing of V_{eff} also improved the NF parameter of the LNA (Liang 2007). Nevertheless, these benefits appear to be strongly frequency dependent. The weak inversion polarisation is only applicable for low or medium frequency designs (for instance, sub 5 GHz for CMOS 0.18um), but a near threshold overdrive voltage could improve the amplifier performance.

6.3.4 *Current Reuse*

The current reuse is another technique that has to be taken under consideration for the lowering of the power consumption of any design. Its benefits have been largely explained along this book (see chapter 5). However, due to the effect of the inclusion of any PMOS transistor for the frequency behaviour and the NF performance, this technique hasto be considered only if the frequency of operation (and bandwidth) allows it.

6.3.5 *Impedances Matching*

With regard to the input and output 50Ω impedance matching, parasitics have an enormous influence. Both the input transistor's parasitic (basically C_{gs}), and the on chip input inductor normally used for that goal match Z_{IN}. From a noise level perspective, the quality factor of this network isa key factor to its success. On the other hand, the $Z_{OUT}$is strongly dependant on the load. In the case of the LC-tank as load, a higher gain and a lower NF at the proximity of the LC frequency of resonance, for a given current consumption.

This load also filters the adjacent frequency bands, avoiding any saturation from out of band tones. Furthermore, this capacitance helps decoupling the DC signals when the LNA is connected to a mixer, which is almost always the case.

6.3.6 *Cascode*

The use of cascode transistors has been theoretically explained in the previous section of this chapter. This added stage (in CG configuration) can be included with a resistive load or withsource degeneration. In that case, an LC tank load would be easily configured using the gate-drain capacitor of this transistor with the upper degeneration inductor. Its main benefit is the improvement of the reverse isolation for a given current consumption, which has positive impact lowering of the influence between the output and the input implied two main advantages:

- Minimization of the CS transistor's C_{gs} disturbance on the frequency and the output gain. This capacitance influence could be worsened by the miller effect, whose effect is minimized by the cascode.
- Improvement of the stability of the LNA. The cascode avoided the interaction between the LC tank and the inductors of the input, if used for input impedance matching. The stability assurance also supposes current consumption saving, as the stability would be otherwise reaches through a higher gain, directly extracted from a higher drain current in the core amplification stage.

6.4 Low-Power LNA Design Examples

This section describes in detail the design of two different low-power LNA examples, regarding its circuit design considerations, layout implementation, the test setup used for the measurement of their main parameters, the characterization results, and finally a short discussion on such results.

- The first example is a low-power shunt-series feedback-based wideband LNA for the DVB-T and DVB-H standards (DVB-T_EN300744), (DVB-H_EN302304).
- The second example shows another low-power LNA for higher frequency applications in the 5GHz U-NII frequency band, such as WLAN (IEEE99), (IEEE03).

6.4.1 Example 1: Low-Power LNA for DVB-T/H

This subsection includes the description of a low-power LNA for DVB-T/H. It comprises circuit design issues, layout implementation considerations, the description of the test setup for the measurement of its main parameters, the characterization results as well as a brief discussion on such results with regard to the stateoftheart.

6.4.1.1 Circuit Design

The DVB-H standard demands a high bandwidth and high linearity performance, which appear to be the most notable requirements of the LNA described within this example. Among the wideband amplifier architectures, the resistive shunt-feedback (Lee 1998) has been extensively used due to its superior broadband characteristics, and it has been the solution implemented in this case.

However, this topology has an important trade-off between noise figure and -3dB bandwidth (also wideband input matching). The -3dB bandwidth of such amplifier architecture is given by the following expression:

$$\omega_{-3dB} = \frac{1+A_v}{R_s\left[C_{gs} + (1+A_v)C_{gd}\right]} \qquad \text{Eq. 6.29}$$

where A_v is the open-loop gain of the amplifier, R_s is the shunt-feedback resistor, and C_{gs} and C_{gd} are the parasitic capacitances of the input transistor. As it can be deducted from Eq. 6.29, higher A_v and smaller R_s lead to wider -3dB bandwidth. However, smaller $R_s$is not a desirable approach due to noise figure degradation. Therefore, A_v can be increased for a wider bandwidth. However, higher A_v requires more DC current consumption in CMOS technology, due to poor transconductance.

In this example, this topology has been implemented and optimized for a wideband input matching and low noise figure. A second stage has been added to extend the -3dB bandwidth. This way, high gain, low noise figure and wide bandwidth are obtained simultaneously.

Fig. 6.20 shows the schematic of the two-stage LNA implemented in this work. The first stage is based on the shunt-series feedback for a wideband input matching and low noise figure. The second stage extends the -3dB bandwidth of the overall amplifier. A third stage consisting on a source follower buffer has been added to ensure suitable matching with the characterization equipment.

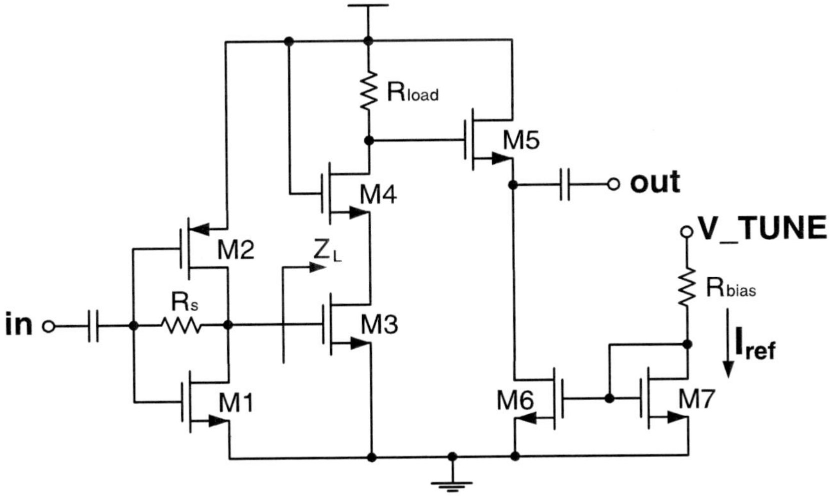

Fig. 6.20 LNA for DVB-T/H circuit simplified schematic

To reduce the power consumption, the current reuse technique has been used for the implementation of the first stage of the amplifier. This way, a NMOS and PMOS transistors pair (M1 and M2 respectively) have been stacked as two amplifying devices. This technique enhances the overall transconductance of the compound, allowing higher R_s for the given -3dB bandwidth, leading to a lower noise figure. The extra capacitance introduced by M2 has minor impact on the bandwidth since the impedance Z_L contributes comparable parasitic capacitance. Both devices are 150µm wide and 0.35µm long. R_s is a 1200Ω poly resistor. This stage drives a current of 6.75mA.

The second stage compensates for the gain roll-off of the first stage and also provides good flatness over the frequency band of interest. It consists on a common source amplifier (M3) with a cascode device (M4). The result accomplished by this device is twofold; it better isolates the output of the amplifier from its input, improving its stability and also avoids the Miller effect on device M3, and so the bandwidth of the amplifier is widened. Both devices have been implemented width the minimum length; conversely, the widths are 20µm and 50µm for M3 and M4 respectively. This stage has been loaded with a 900Ω poly resistor to achieve a better wideband performance and higher gain. The current consumption through this stage is 920µA. Therefore, the total current consumption of the core of the amplifier is 7.67 mA from a 2.8 V power supply, that is, 21.3mW.

A third stage has been also implemented. It consists in a source follower-based output buffer, purely for characterisation purposes in order to drive the 50 Ω input impedance of the measurement equipment. It contributes 2mA to the current consumption of the amplifier.

6.4.1.2 Layout Implementation

Fig. 6.21 shows the layout of the LNA core. All the RF transistors have been implemented exploiting multi-finger configurations. In order to reduce the gate resistance, all the gates have been connected by metal tracks, leading to a reduction in the noise figure. All active devices have been ground-shielded for a better performance against the noise coupled through the substrate for maximum performance. As can be observed inFig. 6.21, the surrounding area has also been shielded with contacts from substrate to ground.

In order to de-couple the noise coming from the power supply, as much capacitance as possible has been implemented from V_{DD} to ground (85pF approx.), by means of:

Tracking wide V_{DD} paths. This way, the parasitic capacitance from the track to the substrate directly adds to the V_{DD}-to-ground capacitance.

Laying out as many capacitors as possible from the V_{DD} paths to ground. For the on-wafer measurement of the LNA, the free space left within the pads and the mixer core has been filled by capacitors (see Fig. 6.22).

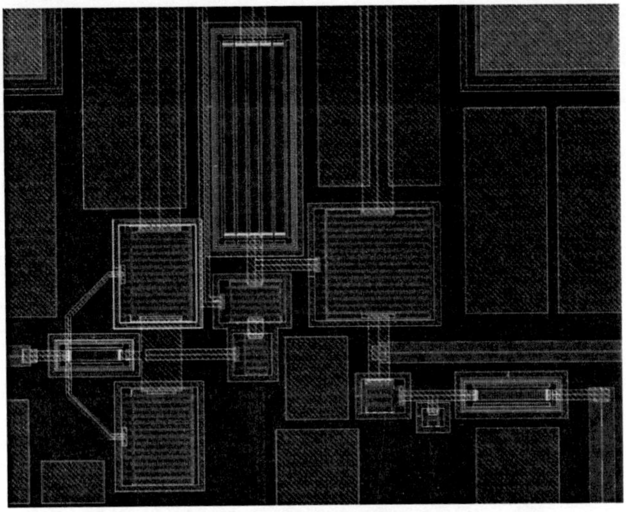

Fig. 6.21 Layout of the LNA for DVB-T/H core

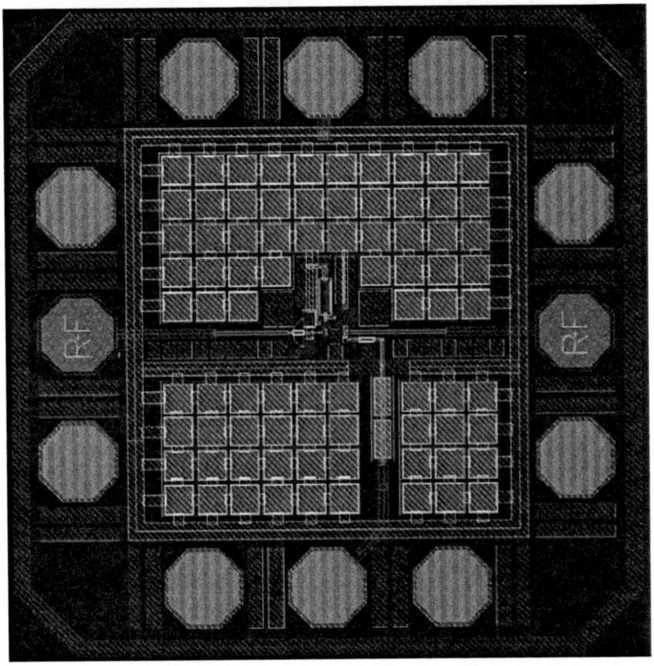

Fig. 6.22 LNA for DVB-T/H layout implementation

6.4.1.3 LNA Test Setup and Characterization Results

The LNA has been measured through an on-wafer probe station. Its die photograph is shown in Fig. 6.23. A detail view of the LNA core is displayed in Fig. 6.24.

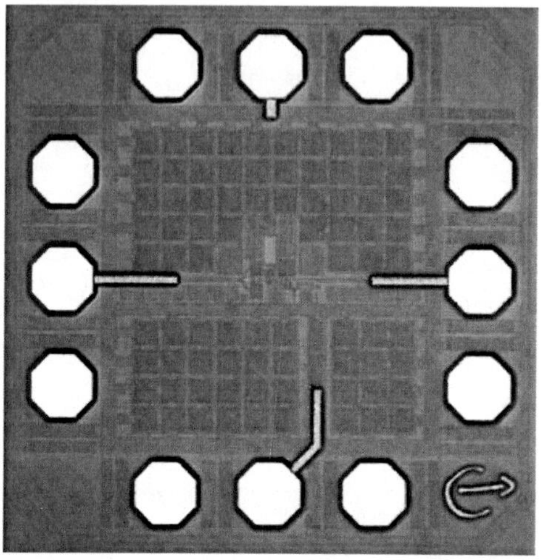

Fig. 6.23 LNA die photograph

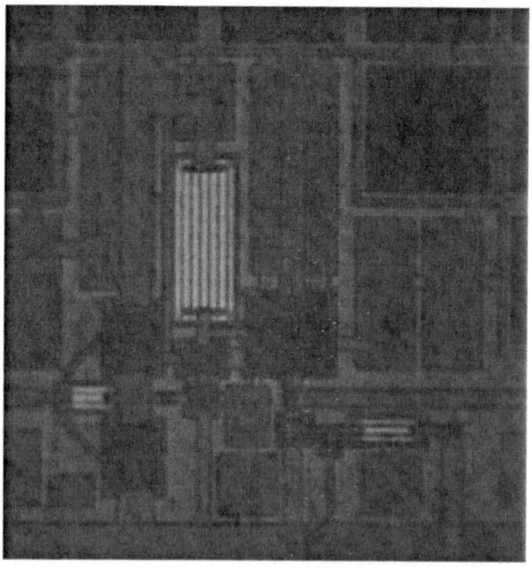

Fig. 6.24 LNA core detail microphotograph

The measurement setup for the S-parameters test is depicted in Fig. 6.25. Since both the input and output of the LNA are matched to 50 Ω, the power gain has been characterized by means of the S21 parameter. The input matching of the LNA (S11) and its gain are depicted in Fig. 6.26 and Fig. 6.27 respectively.

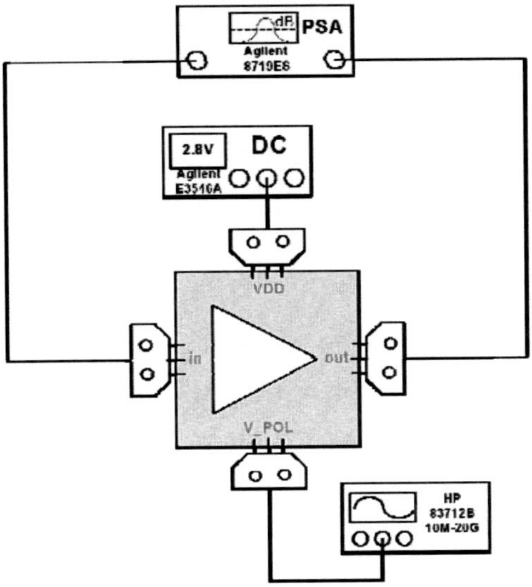

Fig. 6.25 LNA S-parameters measurement setup

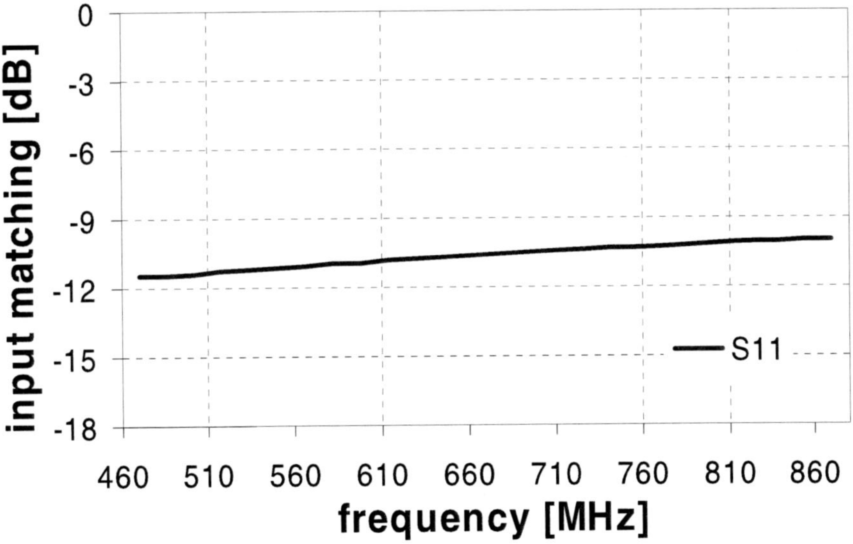

Fig. 6.26 LNA input matching within the UHF band

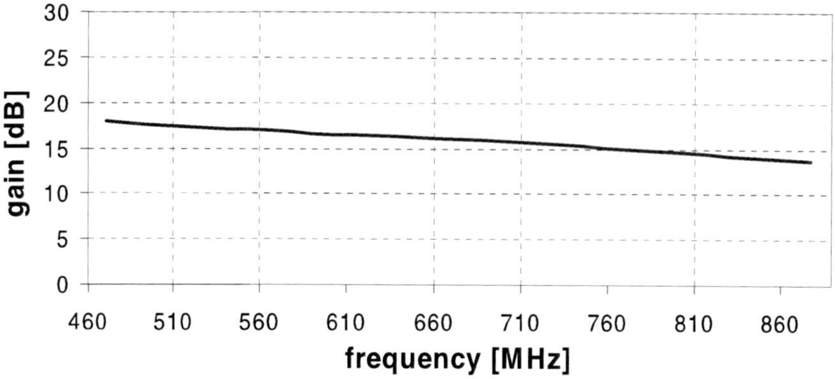

Fig. 6.27 LNA gain in the UHF band

The spectrum analyzer (Agilent E4402B) allows measuring the noise figure with the noise figure option 219 and the noise source 346A. The test setup is shown in Fig. 6.28. The results can be observed in Fig. 6.29.

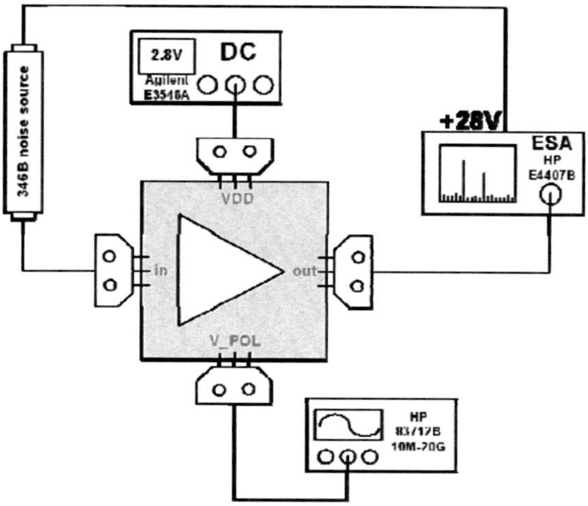

Fig. 6.28 LNA noise figure and gain measurement setup

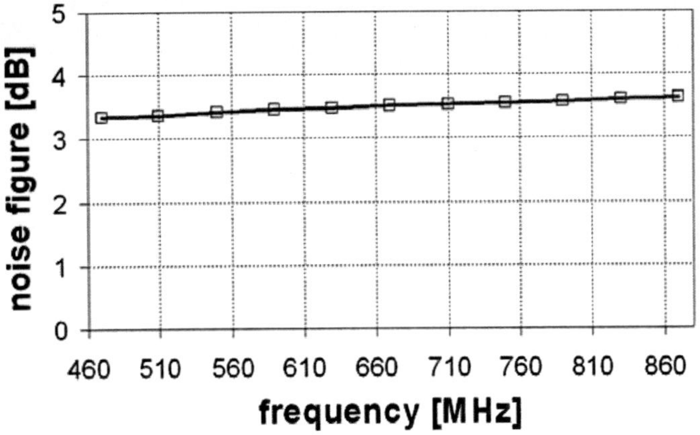

Fig. 6.29 LNA noise figure in the UHF band

Regarding the linearity performance, a two-tone test has been carried out to measure the input-referred third order intercept point (IIP3) of the LNA. The test setup is depicted in Fig. 6.30, while the measured results can be observed in Fig. 6.31.

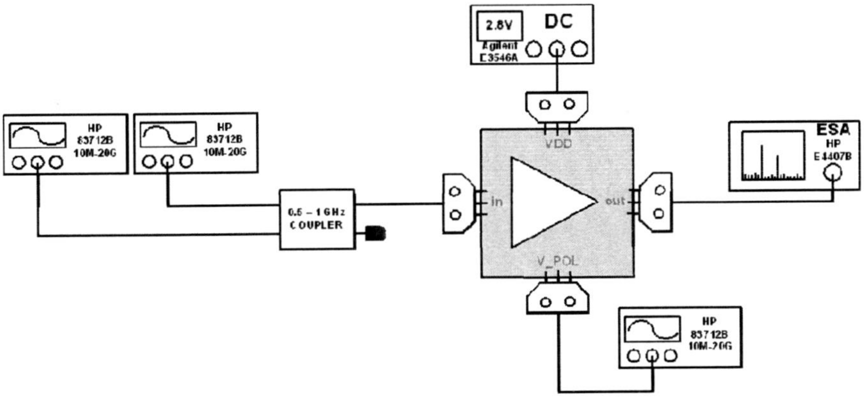

Fig. 6.30 LNA IIP3 measurement setup

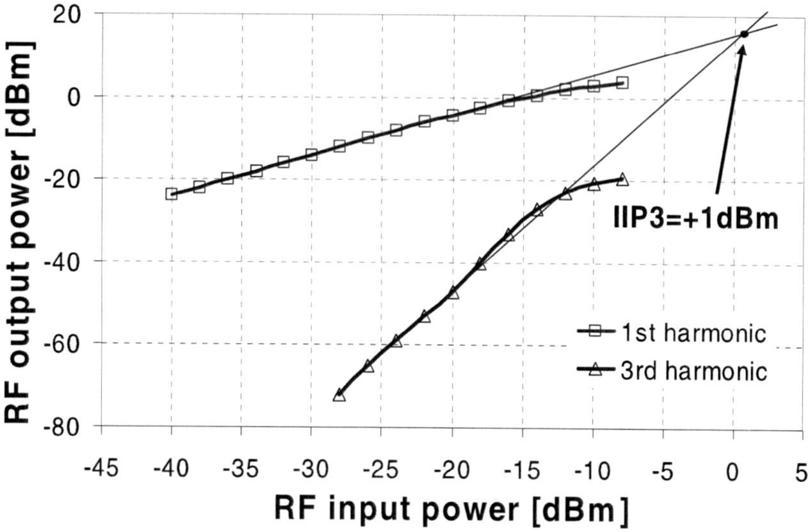

Fig. 6.31 LNA third order intercept point

Finally, Table 6.4 summarizes the measured results of the low noise amplifier.

Table 6.4 LNA measurement results summary

Parameter	Unit	Specification	Measurement
Input matching (S11)	dB	<-10	<-10
Gain	dB	20	16
Reverse Isolation (S12)	dB	Max	-74
NF	dB	2.5	3.4
IIP3	dBm	0	+1
Powerconsumption	mW	min	25

6.4.1.4 Discussion

The results presented in the previous subsection have provided the following performance for a current consumption of 9mA:

Input impedance matching: S11<-10dB
Power gain: S21>16dB
Input third order intercept point: IIP3=+1dBm
Noise Figure: NF=3.4dB

As it can be observed in Fig. 6.26, the input matching is flat within the whole band of interest (470MHz – 862MHz). The S11 parameter is lower than the specification (S11<-10dB). However, post-layout simulations predicted the same behaviour of this parameter but with a minimum value 1dB lower than the measured one. This effect can be explained by the change in the circuit input impedance due to the increase of the measured current consumption of the LNA. Nevertheless, the specification is fulfilled in terms of input impedance matching.

On the other hand, Fig. 6.27 shows the power gain within the frequency band of interest. It can be observed that the measured gain is lower than the specification. It is mainly due to higher parasitic resistance than predicted in the signal path (especially from the input pad to the gates of the input transistors (M1 and M2 in Fig. 6.20) and from the output of the circuit to the output pad. To increase the flatness of S21 (and hence the bandwidth of the amplifier) an inductor could be implemented between devices M3 and M4, but at the cost of higher NF due to the parasitic series resistance of the inductor and higher chip area.

The noise figure is depicted in Fig. 6.29. Its frequency response varies from 3.2dB to 3.6dB at 870MHz. The maximum value of NF is 1.1dB higher than the specification (see Table 6.4). This is due to the parasitic series resistance of the input path, which directly translates into an increase of the gate resistance of the input transistors (M1 and M2 in Fig. 6.20). Other possible cause of the NF deterioration is the reduction of the value of the feedback resistor R_s.

Finally, the linearity performance of the amplifier has been characterized by means of the third order intercept point (IIP3), which has been depicted in Fig. 6.31. Measured at 666MHz, the amplifier exhibits an IP3 of +1dBm referred to the input, which fulfils the specification for this parameter. The gate voltage of devices M1 and M2 has been set just above the threshold voltage. The reduction of the overdrive voltage increases the gain per current unit, lowering the distortion and hence improving the linearity.

Shown in Table 6.5 is a collection of the most representative broadband LNAs of the state-of-the-art. These LNAs are compared with the one described in this section in terms of its most relevant parameters:

- Frequency of operation
- Input matching (S11)
- Power gain
- Noise Figure
- Input referred third order intercept point (IIP3)
- Power consumption

It is difficult to compare the performance of the LNA reported in this work with other LNAs due to the lack of specific DVB-H LNAs in the literature. However, such a comparison can be carried out taking into account other devices working in the same frequency band, and fabricated in standard low-cost technologies.

Regarding power gain, (Janssens et al. 1997) exhibits the highest value at the cost of the lowest IIP3. In addition, the power consumption is also the highest. Therefore, the LNA of this work represents the best performance in terms of power gain taking into account the rest or parameters.

The LNAs presenting lower NF exhibit also lower gain values, with the exception of (Janssens et al. 1997) but, as explained before, at the cost of poor values of both linearity and power consumption.

Table 6.5 State-of-the-art broadband LNAs

[REF]	frequency (GHz)	gain (dB)	NF (dB)	IIP3 (dBm)	powerconsumption (mW)	technology (μm)
(Antoine et al. 2005)	0.47-0.862	14 (Av)	8	+12	---	0.35 SiGeBiCMOS
(Zhang and Kinget 2005)	0.2-1	12	2.6	-10	8.9	0.5 SiGeBiCMOS
(Bruccoleri et al. 2004)	0.25-1.1	13.7	2	0	35	0.25 CMOS
(Janssens et al. 1997)	0.4-1.18	24	2.3	-14.2	35	0.4 CMOS
(Kim et al. 2006)	0.47-0.86	10	5.7	+13	5.2	0.18 CMOS
(Bruccoleri et al. 2002)	0.02-1.6	13.7	1.9-2.4	0	35	0.25 CMOS
(Janssens et al. 98)	0-0.8	14.8	2.3-3.3	-4.7	10	0.5 CMOS
[thisexample]	**0-0.9**	**16**	**3.4**	**+1**	**25**	**0.35 CMOS**

(Antoine et al. 2005) and (Kim et al. 2006) present the highest values of IIP3, using HBTs and 0.18μm CMOS transistors respectively. Furthermore, (Kim et al. 2006) shows the lowest power consumption, but compromising both gain and noise figure, as does (Antoine et al. 2005). Post-layout simulations have shown that the IIP3 of the LNA presented in this work would increase 10dB without the output buffer. With +11dBm, this LNA would represent the best performance in terms of IIP3 taking into account the gain and noise figure.

6.4.2 Example 2: Low-Power LNA for the 5GHz U-NII Band

This subsection includes the description of a low-power LNA for the 5 GHz U-NII band, centred at 5.5 GHz. It comprises circuit design issues, layout implementation considerations, the description of the test setup for the measurement of its main parameters and the characterization results as well as a brief discussion on such results with regard to the state of the art.

6.4.2.1 Circuit Design

As a second example, designed for the 5 GHz U-NII band, here is presented a double-balanced LNA. Even if the gain typically decreases and both NF and power consumption increase, the doubly balanced configuration has been adopted in this design (see Fig. 6.32). The strong rejection of the common mode signals is necessary in front-end balanced architectures with (ideally) identical paths for both I/Q signals.

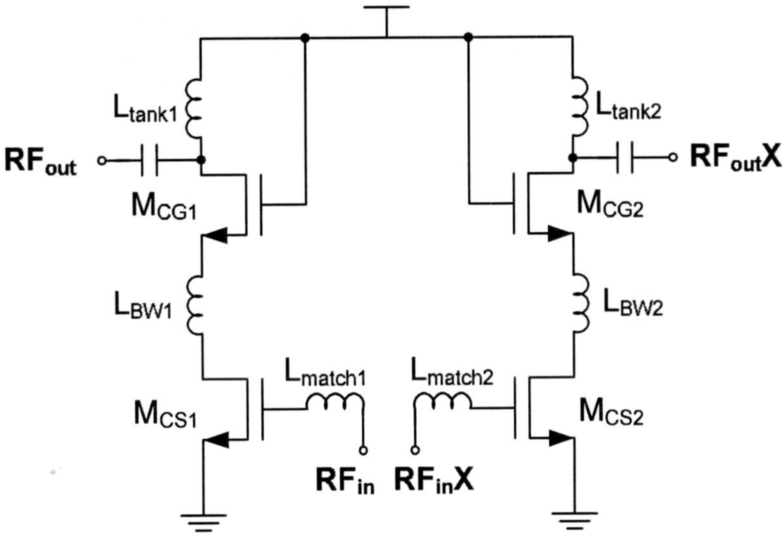

Fig. 6.32 LNA for U-NII frequency band simplified schematic

The design consideration presented in the third section of his chapter and focused on the current consumption optimization are here considered and particularized for the design conditions of this example. Some have been taken and other discarded. The final solution adopted can be observed in the schematic of Fig. 6.32.

The inductive degeneration is used to decouple the Dynamic Range from the power consumption, which has a positive effect on the linearity for a given current consumption. Unfortunately, for the 5 GHz U-NII band, the theoretical values for these degeneration devices may result in inductors in the same order of magnitude than the parasitics. In addition, the gain losses need to be redressed by this linearity extension, so this technique showed no benefits within the present application.

As explained before (see chapter 5), the benefits of the subthreshold polarization appear to be strongly frequency dependent. The weak inversion polarisation is not applicable in the 5 GHz U-NII band, but a near threshold overdrive voltage could improve the amplifier performance.

The current reuse is a key technique that has to be taken under consideration for the lowering of the power consumption of any design. However, due to the effect of the inclusion of any PMOS transistor for the frequency behaviour and the NF performance, and considering the high frequency demand of the 5 GHz U-NII band, this technique has not been adopted for this example.

On the other hand, the core amplification is build with a CS stage. An inductor (L_{match}) connected to the gate of the core amplification transistor has been combined with this transistor's parasitic (C_{gs}) to obtain a matched Z_{IN}. From a noise level perspective, the quality factor of this network has been key to its success.

To avoid an extra current consumption to reach the desirable parameters (Gain, noise figure, linearity, etc.), an LC tank load has been chosen. This resonating output stage provides a higher gain and a lower NF at the proximity of the LC frequency of resonance. This load also filters the adjacent frequency bands, avoiding any saturation from out of band tones. The extracted view simulations of the whole circuit advised the addition of a capacitive series-network at the output. The Z_{OUT} has been strongly dependant on the LC integrated tank (values shown below).

Moreover, a cascode stage with source degeneration (L_{BW}) has been included in the architecture of this LNA. As explained before, the capacitor needed for the LC tank has been obtained from the gate-drain capacitor of these transistors.

But, owing to the wide frequency band of interest, an inductor has been added between the CS and the CG stages. This integrated inductor improves the gain bandwidth at the output (Analui and Hajimiri 2004). It desensitized the load capacitances from the input stage.The CG transistors used in the cascodes have been dimensioned in concordance with the capacitors needed for the LC tanks, but with no disruptions for the output signal properties. The RF optimized transistors selected had minimum length and 50 μm width in 10 fingers. Both these transistors and the CS ones have been cautiously integrated and matched in the layout of the chip by the common centroid technique.

From the numerical point of vioew, the V_{eff} of each CS transistor has been calculated as 575 mV. The total power consumption of this LNA is 7.803 mW. With regard to the passive elements, they have been directly obtained from the UMC 0.18 CMOS 1P6M technology libraries:

- L_{IN}: 2.2 nH (6 μm width, 169.08 μm diameter, 2.5 turns)
- $L_{cascode}$: 2.9 nH (10 μm width, 126 μm diameter, 3.5 turns)
- L_{tank}: 2.2 nH (20 μm width, 146.58 μm diameter, 2.5 turns)
- C_{OUT}: 343 pF

6.4.2.2 Layout Implementation

Fig. 6.33 shows the layout view of the LNA described in this example. It can be observed that most of the die area is occupied by the integrated inductors.

Fig. 6.33 Layout of the LNA for U-NII frequency band

6.4.2.3 Test Setup and Characterization Results

A microphotography of the fabricated circuit is shown in Fig 6.34. The results of its characterization are presented in the next subsection. Fig 6.35 exhibits the detail of the CS transistors of the core of this LNA. The 25 fingers per branch have been separated in 5 transistors to compose the common centroid distribution. This assures the symmetry of the balanced configuration.

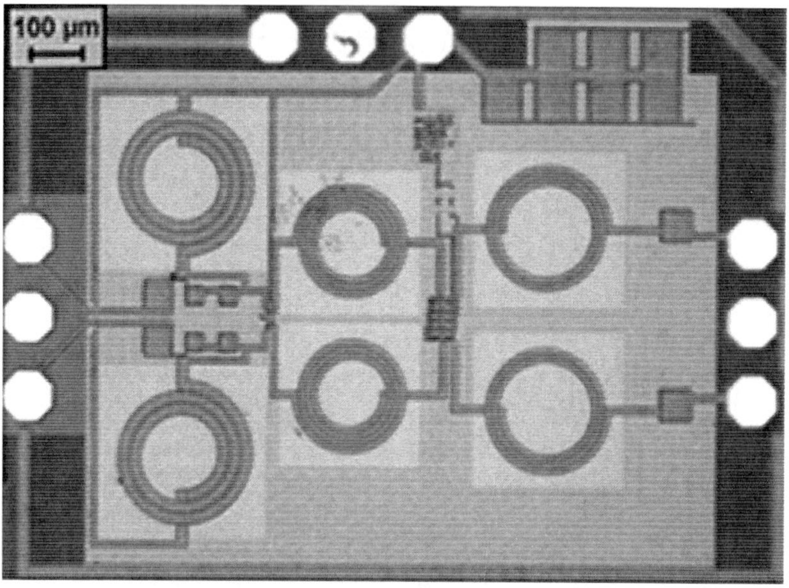

Fig. 6.34 LNA for U-NII band die microphotography

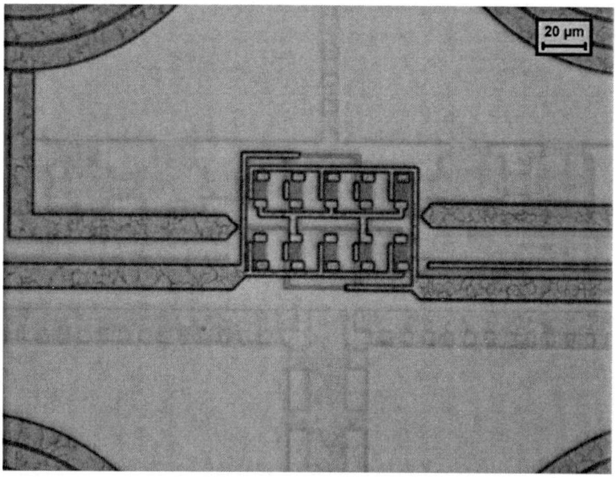

Fig. 6.35 LNA coredetail(commoncentroid) microphotography

The characterization of the LNA has been achieved in three steps. The first one has been the measurement of the S-parameters (measurement setup is shown in Fig. 6.25). From Fig. 6.36 to Fig. 6.39 the results of the measured S11, S12, S21 and S22 are ploted.

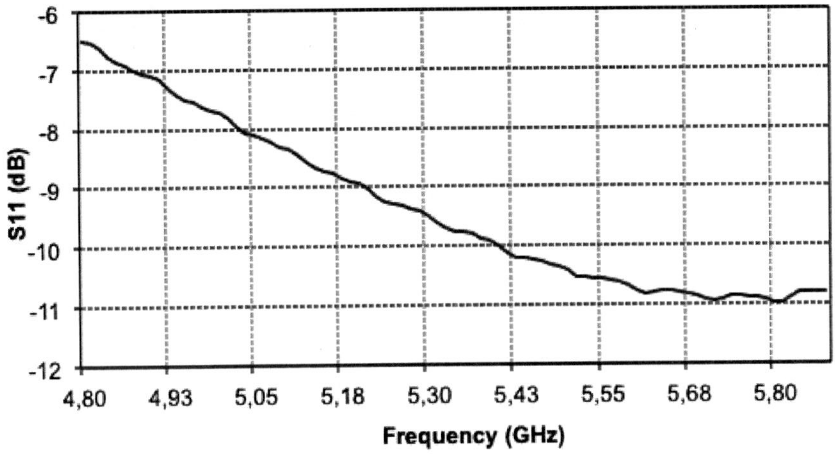

Fig. 6.36 LNA input matching measurement (S11)

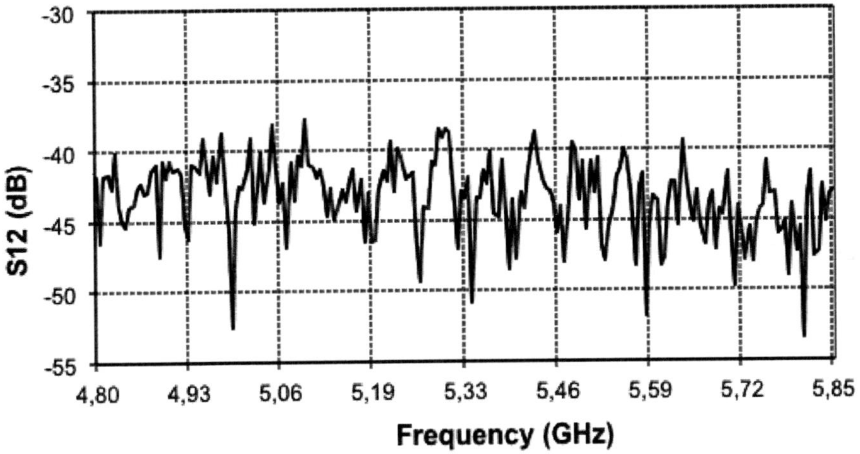

Fig. 6.37 LNA reverse isolation measurement (S12)

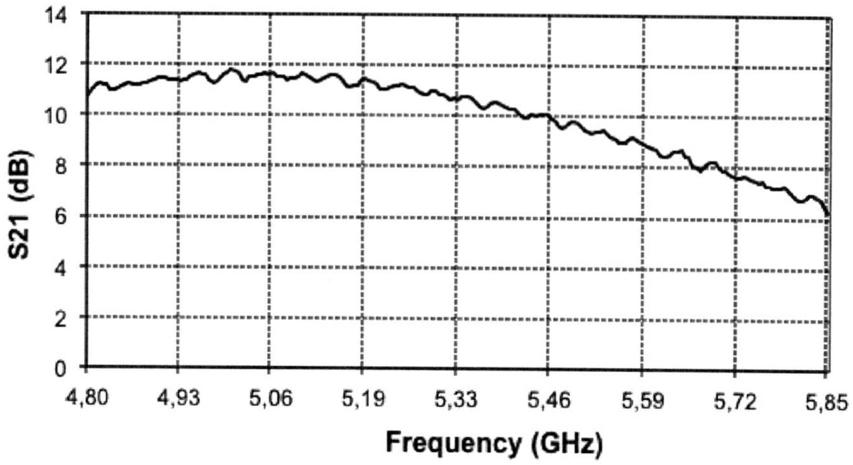

Fig. 6.38 LNA Power gain measurement (S21)

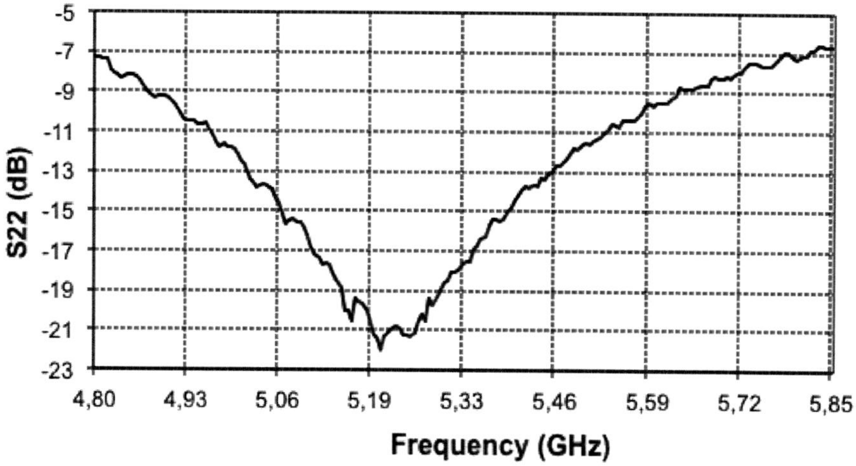

Fig. 6.39 LNA output matching measurement (S22)

The IIP3 measurement setup is a two-tone test, as depicted in Fig. 6.30. Two input tones, at 5.25 GHz and 5.245 GHz respectively, are amplified by the LNA, and the two intermodulation products appear at 5.24 GHz and 5.255 MHz. The result for the IIP3 is shown in Fig. 6.40.

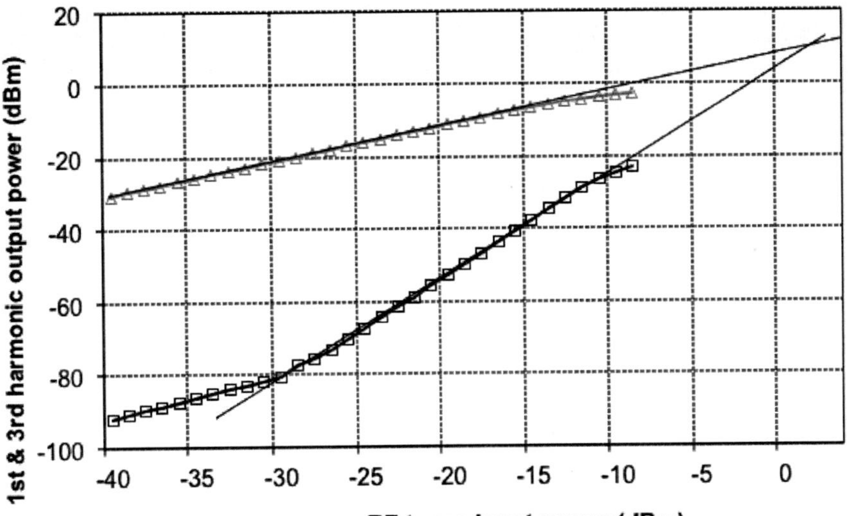

Fig. 6.40 LNA IIP3 measurement

Finally, with the measurement setup shown in Fig. 6.28, the NF characterization is plotted in Fig. 6.41

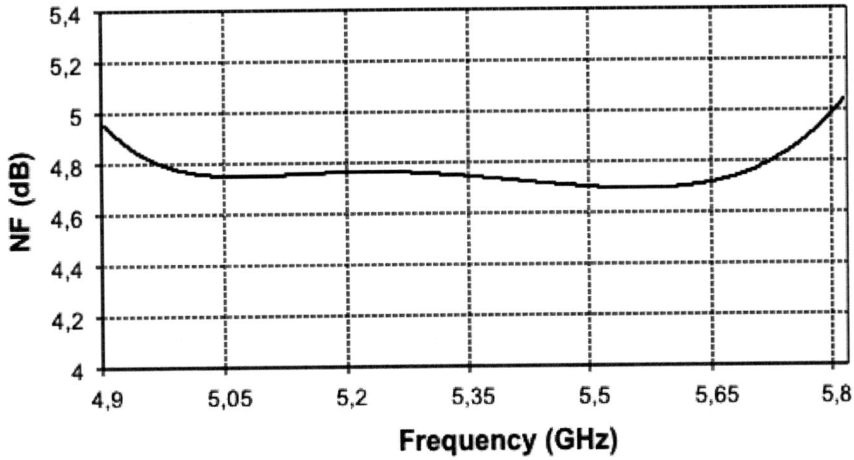

Fig. 6.41 LNA Noise Figure measurement

6.4.2.4 Discussion

The results of the LNA presented above have provided the following performance for a current consumption of 4.335 mA:

- Input and output impedance matching: S11>-11 dB and S22>-21 dB.
- Reverse isolation: S12=-39 dB
- Gain: S21=11.8 dB
- IIP3 measured at 5.25 GHz: +1 dBm
- Noise Figure: 4.7 dB

The bandwidth of the input impedance matched to 50 Ω, shown inFig. 6.36, is higher than the required by the 5 GHz U-NII band (IEEE99). The widening is due to the low Q value of the integrated inductor of the matching network. The width of this L has been minimized (6 μm), so that the bandwidth is maximised.

On the other hand, the quality factor of the output matching network is higher than the Q of the input matching network, as it is exclusively composed of capacitors. The plot of Fig. 6.39 exhibits an appropriate value of the S22 for the 5 GHz U-NII bands I and II, whereas band III adaptation is harmed by values higher than -9 dB. The inclusion of an inductor in this output network can benefit the widening of the S22 parameter, to the detriment of the chip area and a higher complexity of this matching network.

The S12, shown in Fig. 6.37, illustrates the reverse isolation, which reaches a constant -39 dB level. The cascode included in the architecture has increased the result of this parameter. The isolation of this stage is summed to the one of the mixer, for the front end final isolation (see chapter 3).

The gain, plotted in Fig. 6.38, reaches 11.8 dB, and the 3 dB bandwidth covered by this LNA is extended through 1.2 GHz. This relatively wide bandwidth proves the availability of the inclusion of the integrated inductor between the core of the amplifier and the cascode. Furthermore, the track width of the integrated inductor of the tank has been maximized (20 μm). The high Q of the LC tank has benefited a higher gain.

The linearity is higher than the required. As presented in the design procedure, V_{gs} and V_{ds} of the common source amplifier in the core of this LNA have played a decisive role. The gate polarization of the amplifier just above the threshold voltage has benefited it with high gain per current unit, which has lowered the distortion. On the other hand, V_{ds} has been maximized also to improve the linearity. Although the current reuse technique would have lowered the current consumption of this device, the inclusion of the PMOS transistor needed for this technique would have diminished the V_{ds} of the CS amplifier. This technique is not applicable with such frequency and IIP3 requirements.

The Figure of Merit that provides the most complete contrasting for LNAs in the band of operationhas been obtained from (Liang 2007). This publication differences the Low Noise Amplifiers among the bandwidth of their application. In this case, the 675 MHz width of the band of interest binds the amplifier of this work with the relation exhibited in the following equation:

$$FOM_{LNA}[GHz] = \frac{Gain[abs] * IIP3[mW] * BW[GHz]}{(NF[abs]-1) * Power[mW]} \qquad \text{Eq. 6.30}$$

The LNA showed within this exampleis confronted with other designs extracted from the literature in , where the higher values represent the most complete designs in terms of the FOM of Eq. 6.30.

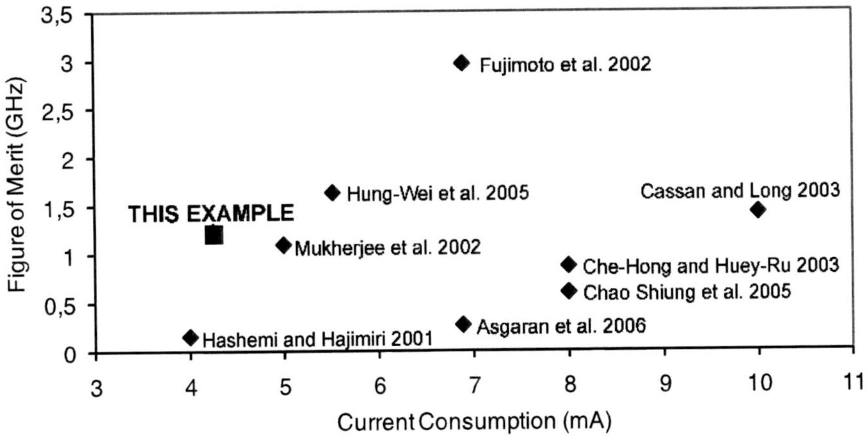

Fig. 6.42 Figure of Merit of the CMOS LNAs working in the 5 GHz U-NII band

The LNA of this work meets the FOM of the amplifiers for similar applications around the 5 GHz U-NII band. Moreover, it exhibits the lower power consumption, 7.65 mW with a current consumption of 4.25 mA. The lower current is performed by the design presented by (Hashemi et al. 2001), with 4 mA. Nevertheless, the performance of this design is far away from the requirements of this example, as the bandwidth does not go beyond 100 MHz.

Nevertheless, the LNA reported by (Fujimoto et al. 2002) presents a substantial improvement in the FOM value, in comparison with the others reported. This is due to a dual gate MOSFET technology fabrication. This non-standard technique benefits the global behaviour with the inclusion of the common source amplifier and the cascode in the same transistor. The dispersions and the parasitic mismatches are minimized in such a delicate block. Thus, the operating frequency of this LNA reached 6.9 GHz and a bandwidth of 3.1 GHz. In addition the gate resistance of the transistors used in this referenced publication is reduced thanks to a low resistance salicided gate material, which improves the NF performance.

The LNA presented as this second example is, hence, in the same level than the best standard designs previously considered in the state of the art, but with the lower power consumption reported.

References

(Abidi et al. 2000) Abidi, et al.: Power-conscious design of wireless circuits and systems. Proceedings of the IEEE 88(10) (2000)

(Analui and Hajimiri 2004) Analui, Hajimiri: Bandwidth enhancement for transimpedance amplifiers. IEEE Journal of Solid-State Circuits 39(8) (2004)

(Antoine et al. 2005) Antoine, P., et al.: A direct-conversion receiver for DVB-H. IEEE Journal of Solid-State Circuits 40(12), 2536–2546 (2005)

(Asgaran et al. 2006) Asgaran, D., et al.: A 4-mW monolithic CMOS LNA at 5.7GHz with the gate resistance used for input matching. IEEE Microwave and Wireless Components Letters 16(4) (2006)

(Bruccoleri et al. 2002) Bruccoleri, F., et al.: Noise cancelling in wideband CMOS LANs. In: 2002 IEEE International Solid-State Circuits Conference, ISSCC. Digest of Technical Papers, pp. 330–533 (2002)

(Bruccoleri et al. 2004) Bruccoleri, F., et al.: Wide-band CMOS low-noise amplifier exploiting thermal noise canceling. IEEE Journal of Solid-State Circuits 39(2), 275–282 (2004)

(Cassan and Long 2003) Cassan, D.J., Long, J.R.: A 1-V transformer-feedback low-noise amplifier for 5-GHz wireless LAN in 0.18-um CMOS. IEEE Journal of Solid-State Circuits 38(3) (2003)

(Caverly 2007) Caverly, R.: CMOS RFIC Design Principles. Artech House Publishers, Boston (2007)

(Chao-Shiun et al. 2005) Wang, C.-S., et al.: A multi-band multi-standard RF front-end IEEE 802.16a for IEEE 802.16a and IEEE 802.11 a/b/g applications. In: IEEE International Symposium on Circuits and Systems, ISCAS 2005, May 23-26, vol. 4, pp. 3974–3977 (2005)

(Che-Hong and Huey-Ru 2003) Che-Hong, Huey-Ru: A 5.7-GHz 0.18um CMOS gain-controlled differential LNA with current reuse for WLAN receiver. IEEE Microwave and Wireless Components Letters 13(12) (2003)

(DVB-H_EN302304) DVB-H_EN302304, Transmission System for Handheld Terminals, ETSI standard

(DVB-T_EN300744) DVB-T_EN300744, Framing Structure, Channel Coding and Modulation for Digital Terrestrial Television, ETSI standard

(Fujimoto et al. 2002) Fujimoto, K., et al.: A 7-GHz 1.8-dB NF CMOS low-noise amplifier. IEEE Journal of Solid-State Circuits 37(7) (2002)

(Hashemi and Hajimiri 2001) Hashemi, H., Hajimiri, A.: Concurrent dual-band CMOS low noise amplifiers and receiver architectures. In: 2001 Symposium on VLSI Circuits, Digest of Technical Papers, pp. 247–250 (2001)

(Horenstein 1995) Horenstein, M.N.: Microelectronic Circuit and Devices. Prentice-Hall, Englewood Cliffs (1995)

(Hung-Wei et al. 2005) Hung-Wei, S., et al.: A 2.17-dB NF 5-GHz-band monolithic CMOS LNA with 10-mW DC power consumption. IEEE Transactions on Microwave Theory and Techniques 53(3) (2005)

(IEEE99) IEEE, WLAN 802.11a: Wireless LAN Medium Access Control (MAC) and Physical Layer (PHY) Specifications. High-speed Physical Layer in the 5 GHz Band (1999)

(IEEE03) IEEE, Part 11: Wireless LAN Medium Access Control (MAC) and Physical Layer (PHY) specifications Amendment 4: Further Higher Data Rate Extension in the 2.4 GHz Band, IEEE, SA Standard Board (2003)

(Janssens et al. 1997) Janssens, J., et al.: A 2.7-V CMOS broad-bandlow-noise amplifier. In: 1997 Symp. VLSI Circuits Dig. Tech. Papers, pp. 87–88 (1997)

(Janssens et al. 1998) Janssens, J., et al.: A 10 mWinductorless, broadband CMOS low noise amplifier for 900 MHz wireless communications. In: Proceedings of the IEEE 1998 Custom Integrated Circuits Conference, May 11-14, pp. 75–78 (1998)

(Kim et al. 2006) Kim, T.W., Kim, B.: A 13-dB IIP3 improved low-power CMOS RF programmable gain amplifier using differential circuit transconductance linearization for various terrestrial mobile D-TV applications. IEEE Journal of Solid-State Circuits 41(4), 945–953 (2006)

(Lee 1998) Lee, T.: The Design of CMOS Radio Frequency Integrated Circuits. Cambridge University Press, Cambridge (1998)

(Liang 2007) Liang: The Exploration of Spectrum Monitor Architecture for Cognitive Radio, University Of Southampton, PhD dissertation (2007)

(Mukherjee et al. 2002) Mukherjee, D., et al.: A 5-6 GHz fully-integrated CMOS LNA for a dual-band WLAN receiver. In: IEEE Radio and Wireless Conference, RAWCON 2002, pp. 213–215 (2002)

(Sedra 2009) Sedra, A.S., Smith, K.C.: Microelectronic Circuits. Oxford University Press, Oxford (2009)

(Zhang and Kinget 2005) Zhang, F., Kinget, P.: Low power programmable-gain CMOS distributed LNA for ultra-wideband applications. In: 2005 Symposium on VLSI Circuits, Digest of Technical Papers, June 16-18, pp. 78–81 (2005)

7 Mixer Design

The mixer is one of the key building blocks in any transceiver system as it performs the frequency up/down conversion, especially in the receivers, where weak RF signals are present at the input (section 3.1). In most cases, amplified only by the LNA, the mixer has to guarantee low noise figure (a positive gain helps for this issue), and high linearity, while consuming low-power consumption. As in any RF circuit, its whole performance, taking into account all these parameters, represent a severe trade-off. The aim of this chapter is to analyze the functional aspect of mixers from the low power consumption perspective.

This chapter begins with a brief review of mixer fundamental concepts (section 7.1) and different circuit topologies (section 7.2), as well as a short description of the mixer's main functional parameters. These parameters areanalyzedin depth from the low power consumption perspective in section 7.3, and finally some implemented circuits are presented, analyzed and discussed in section 7.4.

7.1 Mixer Fundamentals

In almost every modern communication transceiver, the information transmitted is converted to different frequencies throughout its analog processing by means of mixing stages(Razavi 1998), (see Fig. 7.1). The basic principle used to perform this frequency conversion is the multiplication of waves in the time domain(Razavi 1998), (Lee 1998).This way, RF and LO signals are multiplied and as a result an IF signal is obtained.

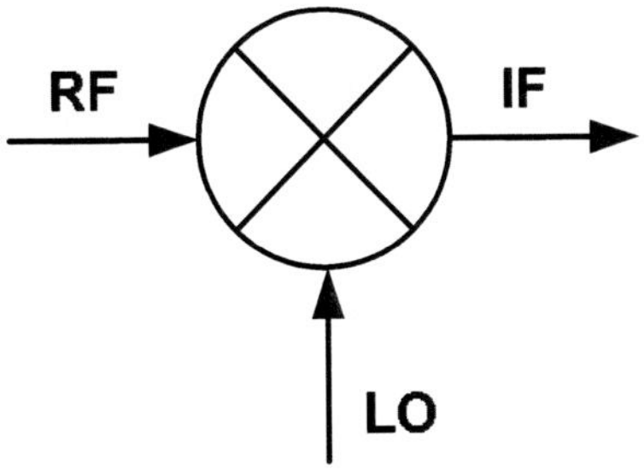

Fig. 7.1 Basic mixer symbol

Taking into consideration the simplest case of two sine waves for both RF and LO, Eq. 7.1 shows the expression obtained for the IF output.

U. Alvarado et al.: Low Power RF Circuit Design in CMOS Tech., LNEE 104, pp. 129–177.
springerlink.com

$$V_{IF}(t) = A\cos(\varpi_{LO}t)B\cos(\varpi_{RF}t) = \frac{AB}{2}\left[\cos(\varpi_{LO} - \varpi_{RF})t + \cos(\varpi_{LO} + \varpi_{RF})t\right] \quad \text{Eq. 7.1}$$

As explained in the next sections of this chapter, many mixers work with a periodic square wave as LO signal, rather than a sinewaveform. In this case, LO expression can be written as follows:

$$V_{LO}(t) = \frac{4}{\pi}\left(\cos(\varpi_{LO}t) - \frac{1}{3}\cos(3\varpi_{LO}t) + \frac{1}{5}\cos(5\varpi_{LO}t) - \ldots\right) \quad \text{Eq. 7.2}$$

Therefore, mixing occurs between LO's odd harmonics and RF signal, leading to a multiple translation of the input frequency range. In general, the mixer generates outputs at a range of frequencies given by $m\varpi_{RF}+n\varpi_{LO}$, where both m and n are integer numbers (Egan 2003), (Coleman 2004). RF and LO components are also expected due to the lack of total isolation between ports. Fig. 7.1 shows the output spectrum of a downconversion mixer. Most of these outputs tones are undesired side-effects of the mixing process, and are called spurs. Of course, a subsequent filtering stage is needed to select the desired frequency-shifted IF signal.

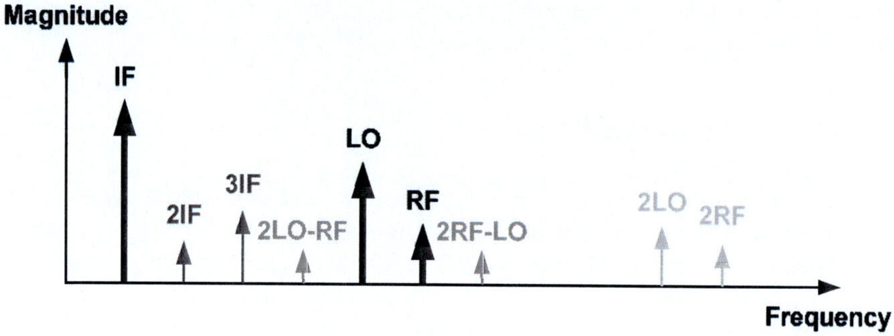

Fig. 7.1 Typical output spectrum of a downconversion mixer

Another important issue of mixers based on multipliers is the image frequency conversion (section 3.1). As depicted in Fig. 7.2, the signal RF-2IF is downconverted directly to the same frequency as the RF. Noise and unwanted signals present at this frequency degrade the system performance. Therefore, special techniques must be applied in order to attenuate this negative effect depending on the architecture selected for the transceiver front-end (section 3.4).

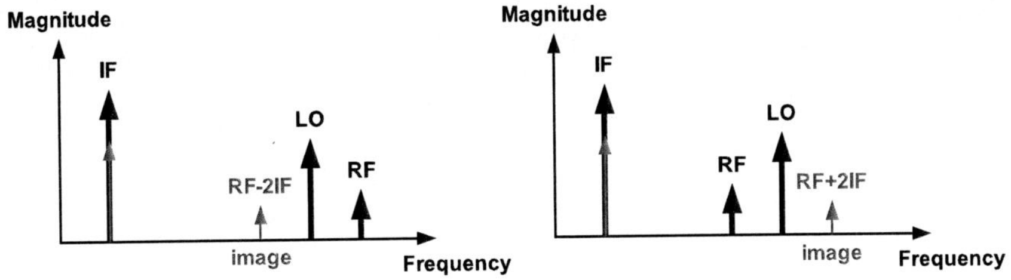

Fig. 7.2 Image frequency downconversion

The main parameters used to compare the performance of mixers are mostly shared among all the RF building blocks presented in this book. They are analyzed in depth from the power consumption perspective in section 7.3, and are briefly outlined through the following subsections.

7.1.1 Conversion Gain / Loss

Conversion Gain/Loss is the ratio of the IF signal level to the input RF signal level, usually expressed in dB. IF and RF are located at different frequency ranges, therefore special calculation/characterization methods must be applied in order to determine this parameter.

The most common formulation for this parameter is the ratio of the *rms* voltage the signal at the output port (IF) to the *rms* voltage of the signal at the input port (RF). If the input and output ports are perfectly matched to a reference impedance (usually 50Ω), voltage conversion gain and power conversion gain are equal.

7.1.2 Linearity

The linearity of mixers is usually characterized using the Third Order Intersection Point (IP3). It is measured by applying two closely spaced input tones at frequencies RF1 and RF2. In the case of a downconversion mixer, third order intermodulation products from the mixing of these tones with the LO occur at frequencies given by (2RF1±RF2)±LO and (2RF2±RF1)±LO, as shown in Fig. 7.3.

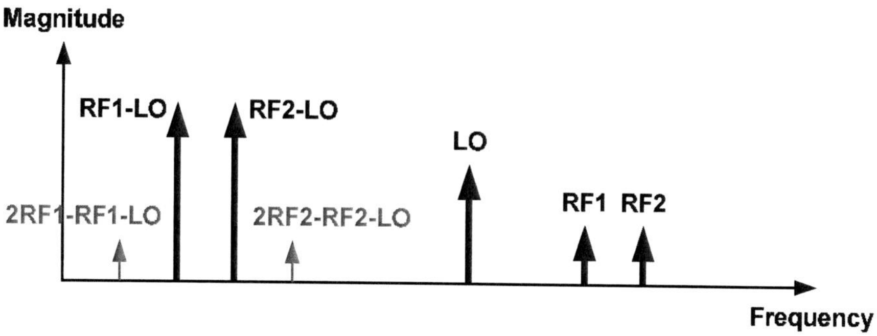

Fig. 7.3 Downconversion of third order IM products

IP3 is defined as the input/output power value that equates the linear extrapolation of the third order intermodulation product and the direct downconverted main harmonic. Fig. 7.4 explains this concept. IIP3 (Input referred IP3) and OIP3 (Output referred IP3) are the input or output referred values of IP3.

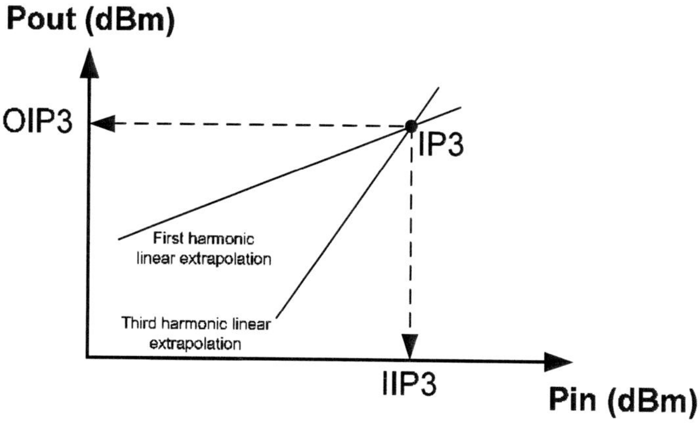

Fig. 7.4 IP3 definition

Another way to describe the linearity performance of a mixer is related to its 1dB compression point (section 3.1). Such a point is defined as the power (in dBm) of the input signal for which the power of the output signal is degraded 1dB below its theoretical linear first order harmonic extrapolation. This concept is depicted in Fig. 7.5:

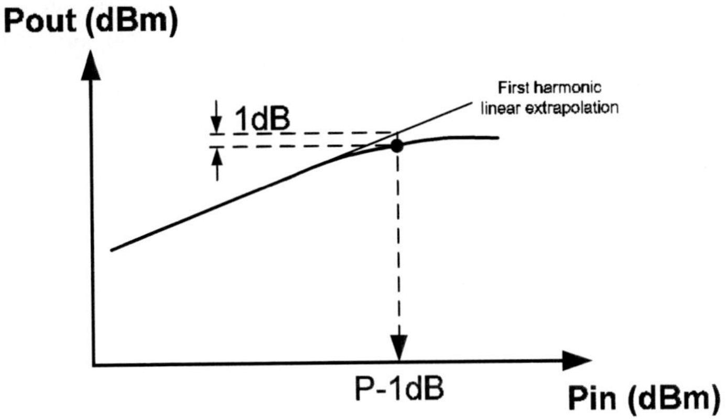

Fig. 7.5 1-dB Compression point definition

The Input referred IP3 and the 1-dB compression point relation can be approximated with the following expression (Razavi 1998):

$$IIP3(dBm) \approx P1dB(dBm) - 9.6 \qquad \text{Eq. 7.3}$$

7.1.3 Noise Figure

Noise Figure (NF) is defined as the ratio of the Signal to Noise Ratio (SNR) at the input compared to the SNR at the output. Depending on the characteristics of the device under study, there are two options:

- Double Sideband (DSB) Noise Figure: Includes noise and signal contributions at both the RF and the image frequencies.
- Single Sideband (SSB) Noise Figure: No image signal is included although image noise is included.

SSN and DSB NFs are related as pointed in Eq. 7.4 (Caverly 2007):

$$NF_{DSB} = NF_{SSB} - 3dB \qquad \text{Eq. 7.4}$$

7.1.4 Impedance Matching and Port Isolation

Impedance matching and port isolation are directly related to the transmission and reflection of the signals (RF, IF and LO) through the different ports. As outlined in section 3.5, transmission from the RF port through the IF port should be maximized for the optimum conversion gain. On the other hand, undesired signal paths have to be minimized by means of proper layout techniques, especially the one related to LO leakage to the RF port. This leakage, in addition to a bad input impedance matching, adds to the reflected signal back through the reverse path of the LNA (Razavi 1998)contributing to the signal re-radiation (section 3.5).

7.2 Mixer Topologies

In this section the most important mixer topologies are summarized, concerning the power consumption classification criterion. This way, active mixers (voltage supply needed for their operation) and passive mixers (supply voltage not needed for their operation) are presented in the following subsections.

7.2.1 Active Mixers

Active mixers are based on the combination of two stages (Lee 1998), (Caverly 2007): A voltage to current converter of the RF incoming signal and an analog multiplier. The basic scheme can be found in Fig. 7.6.

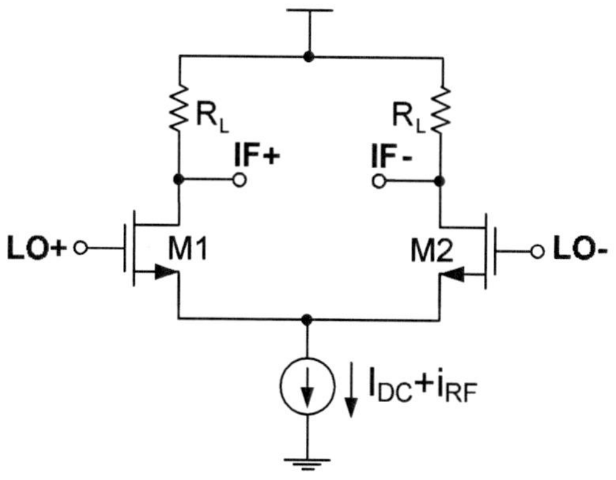

Fig. 7.6 Basic active mixing cell

The core of this scheme is the switching pair. Once an appropriate LO voltage is selected, these two transistors oscillate between saturation and cut-off state, ideally providing a square waveform. Therefore, the current $I_{DC}+i_{RF}$ is alternatively redirected through one of the two R_L branches, provoking the multiplication of LO and RF signals.

The output signal (IF) of the circuit shown in Fig. 7.6 presents a combination of different harmonics of LO signal because of the inclusion of the I_{DC} term in the multiplication. In many occasions, these harmonics are very difficult to remove, as they usually fall in frequencies close to the band of interest. In order to avoid this undesired side effect, double balanced schemes are preferred, as shown in Fig. 7.7.

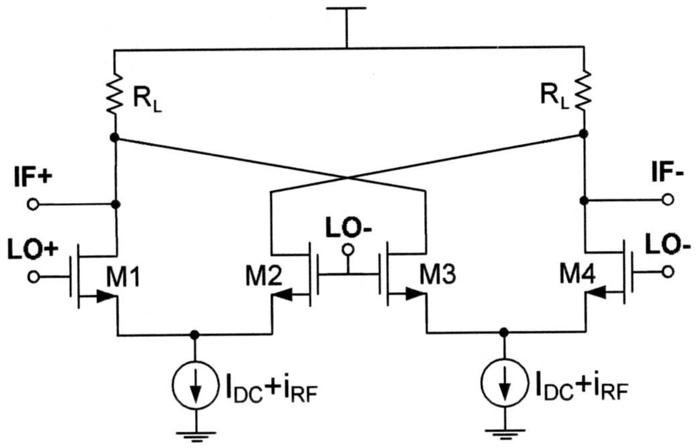

Fig. 7.7 Double balanced active mixing cell

The I_{DC} components are ideally rejected, and thus LO harmonics are not present at IF port. Unfortunately, any non ideality of the switching stage degrades this ideal behavior, contributing to the undesired harmonic content of the output signal.

In both previous cases, a transconductance stage is needed in order to transform the input signal into a proportional i_{RF}. The two most commonly used options are shown in the next subsections.

7.2.1.1 Gilbert Cell Based Active Mixer

The basic form of the Gilbert cell (Gilbert 1968) is shown in Fig. 7.8. In this case, the input stage is implemented using a differential common emitter transconductor. The main characteristics of the mixer (conversion gain, noise figure, IIP3 and power consumption) are crucially determined by its first stage, which is analyzed in depth in section 7.3.

7.2.1.2 Class-AB Active Mixer

This is alternative scheme (Gilbert 1997) offers a solution to several problems that affect the classic Gilbert cell: relatively poor linearity and high input impedance. Although best results are obtained with BJT implementations, CMOS versions can also be laid out. The schematic of this topology is shown in Fig. 7.9.

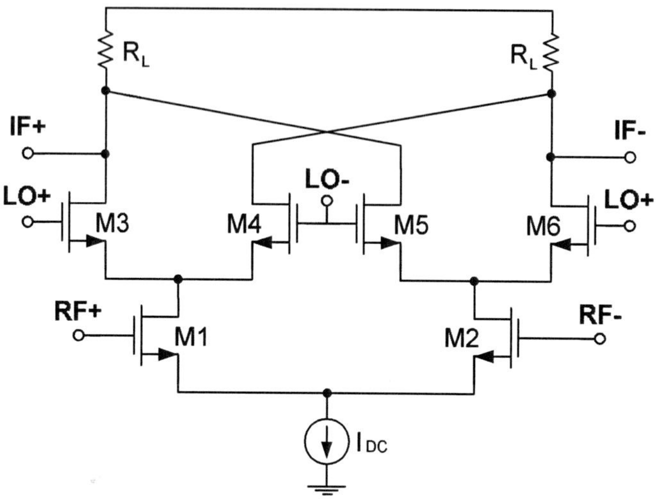

Fig. 7.8 Basic Gilbert cell

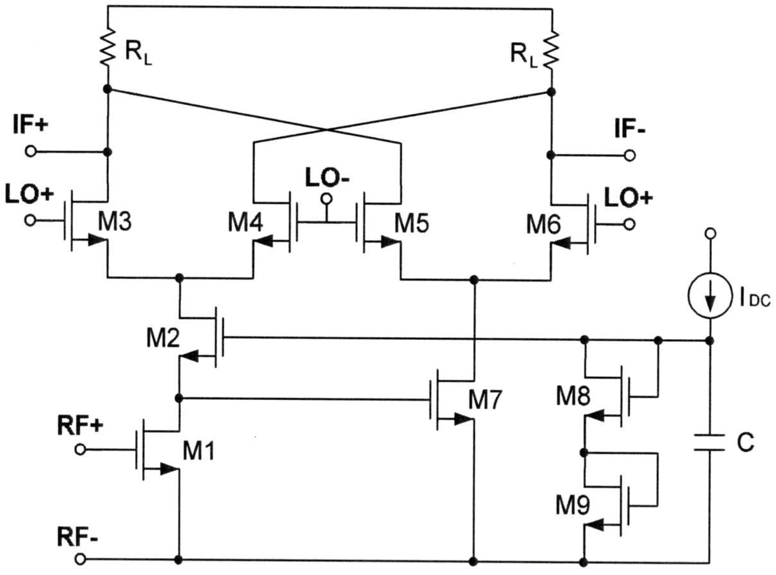

Fig. 7.9 Class-AB mixing cell

In terms of gain conversion, linearity and noise figure, this scheme usually offers better overall results than the classic Gilbert configuration. However, the power consumption is dramatically increased by the additional circuitry needed to implement I_{DC}. For this reason, this circuit is not commonly found in low power applications.

7.2.2 *Passive Mixers*

Passive mixers carry out the frequency conversion by working in the voltage domain, without voltage to current conversion. The saving in power consumption compared to the active solutions is clear, therefore, these appear to be the preferred structures for low power applications. However, this vague conclusion is not very accurate, as the negative conversion gain (conversion loss) then has then to be compensated by other blocks of the system. These general observations are taken into consideration in the low power design constraints section (7.3).

7.2.2.1 Diode Ring Passive Mixer

This is the classic discrete implementation of passive mixers. The four-diode double balanced scheme shown in Fig. 7.10 presents a high linearity over a wide range of frequencies. This would be the obvious choice when linearity is the dominant specification of the circuit.

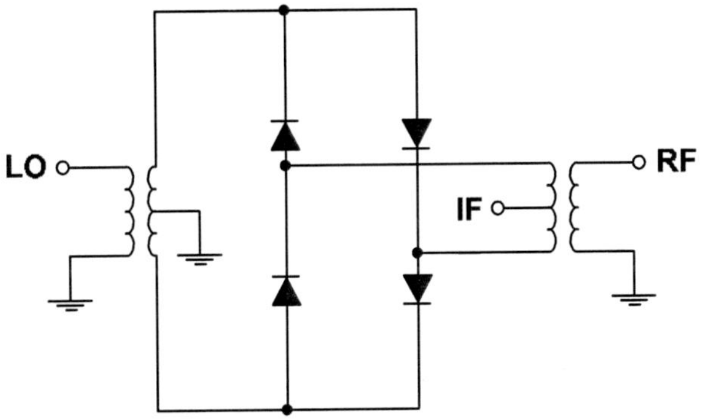

Fig. 7.10 Double balanced diode ring

In this circuit, diodes are used as switches. As the LO signal swings between its positive and negative peaks, the diodes on both columns also switch their state, therefore, at IF port the RF signal appears multiplied by a series of odd harmonics of the LO signal. Because of the lack of amplification, this configuration does not provide conversion gain. The noise added depends mainly on the contribution of forward biased diodes. Finally, a good port-to-port isolation can be achieved.

Despite its simplicity, this architecture has not been widely used in CMOS technologies, due to the lack of low forward voltage drop diodes in the design kits of these fabrication processes. However, this situation is currently changing with the new submicron technologies, and this solution may be suitable for low power applications.

7.2.2.2 Transistor Ring Passive Mixer

Transistor ring (Fig. 7.11) is the straightforward evolution of diode ring, this being the most commonly used approach in integrated CMOS technologies (Maas 1987). As in the previous case, the transistors work as switching devices.

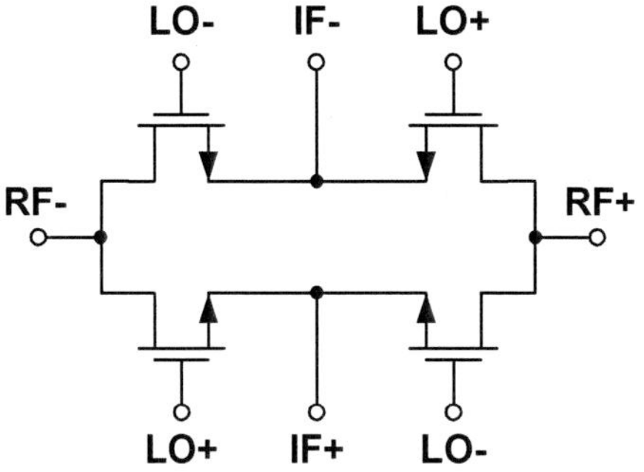

Fig. 7.11 Transistor ring passive mixer

This structure can be compared with the Gilbert cell. In that case, the linearity is somehow compromised by the transconductance input stage. Here, the mixing is performed in the voltage domain, thus providing a high linearity. However, the lack of amplification causes conversion loss and a higher noise figure.

7.2.2.3 Potentiometric Passive Mixer

As frequency of operation increases, the commutation of transistors is less efficient, provoking degradation in the overall performance of the device. One solution for this problem is the use of a slightly different scheme (Crols and Steyaert 1995), in which transistors are biased in the linear region instead of saturation/cut-off. This can be achieved by adopting transistor gates as input port and feeding LO signal by drain and source terminals as shown in Fig. 7.12.

In this configuration, the transistors work as voltage controlled resistors, its variation being proportional to the incoming RF signal. If the drain to source voltage is modulated following LO signal, a multiplication of both signals will be carried out.

The main advantage of this alternative is the high linearity obtained, due to the intrinsic linear operation of transistors. However, the continuous noise contribution of the four transistors raises the value of the noise figure, power consumption is also higher, due to the same reason.

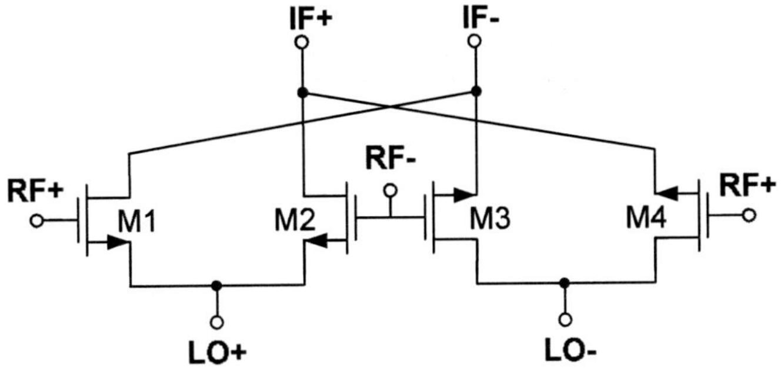

Fig. 7.12 Potentiometric passive mixer

7.3 Mixer Design Constraints

The constraints concerning the power consumption of a mixer have to be faced from two points of view. Firstly, the mixer can be optimized for minimum power consumption. Secondly, the mixer is a block integrated in a down or up conversion system and consequently it affects the blocks connected to him (Razavi1998). While optimizing one of them, the others can be affected, for instance, in a receiver chain, an unwise NF characteristic of a mixer would push the LNA to increase its gain to maintain the receiver chain's NF requirement. This directly means higher power consumption for the LNA. The same reasoning may be considered for other characteristics of the single blocks as they are interconnected and the final result comes from a combination. Care has to be taken from the system point of view.

Considering the mixer itself, the power consumption is related to its functional characteristics: gain, linearity, noise, bandwidth and matching considerations. And moreover, the trends here are also inter-related as presented in the followings subsections.

7.3.1 Gain

The gain of any circuit is always directly related to its power consumption. Moreover, the gain is always sacrificed in benefit of the other characteristics, as the circuit design is a balance game. For constant power consumption, any improvement from the noise, linearity and/or bandwidth point of view requires a gain decreasing.

As previously said in the introduction to the basics of mixers, there is no difference between the voltage conversion gain and the power conversion gain when referenced to 50Ω. The latter is defined as the ratio of the power delivered to the load to the RF power delivered by the source. Furthermore both cases are highly related to the LO power level.

$$ConversionGain = \frac{Power_{Load}}{Power_{Source}} = \left(\frac{I_{Output}}{I_{Input}}\right)^2 \frac{R_{Load}}{R_{source}} \qquad \text{Eq. 7.5}$$

As the mixer is connected within a receiver (or transmitter) chain, R_{Load}, R_{Source} and I_{Input} are fixed by the system, but the mixer has to provide the output current I_{Output}. And, as expected, the Conversion Gain is proportional to it and consequently to the power consumption of the block.

The conversion gain depends on the mixer topology. As seen in section 7.2, the main distinction is active and passive mixers; the former have positive conversion gains, whereas the latter present losses. The design constraints for this parameter for both categories are here developed.

7.3.1.1 Gain in Active Mixers

This kind of topology has several configurations (double balanced, single balanced and unbalanced) but all have a supply source which amplifies the input current in its path to the output load. The conversion gain is here positive. Even if each configuration has its own considerations about gain and power consumption, the base of any active structure in CMOS technology is the switching of the MOS transistors, separated at a transconductance stage, as described in the following subsection, and at a large signal – driven stage.

7.3.1.1.1 Transconductor Stage

Considering an optimum work in the saturation region, the drain current can be calculated with Eq 7.6, considering the channel length modulation:

$$I_D = \frac{1}{2} k' \frac{W}{L} \left(v_{GS} - v_{th}\right)^2 \left(1 - \lambda v_{DS}\right) \qquad \text{Eq. 7.6}$$

This operation is performed to transform the RF signal into a current injected into the LO - driven switching stage. This first stage is called the transconductance stage. From equation Eq 7.6, the transconductance expression in the saturation region is easily obtained by deriving the current with respect tov_{GS}, around the DCV_{GS} operating point.

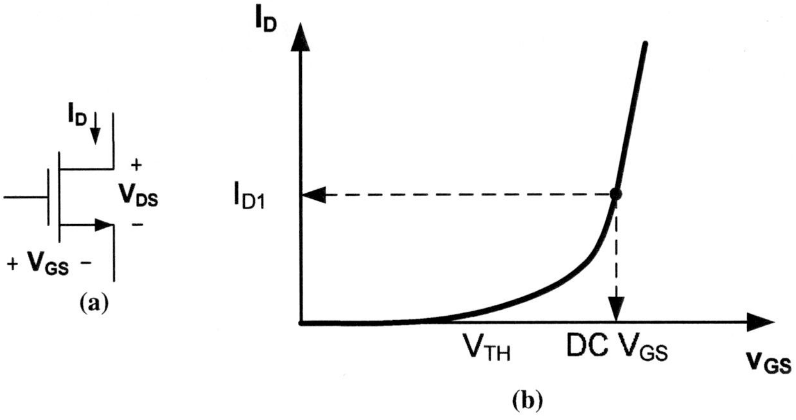

Fig. 7.13 (a) CMOS n-type transistor. (b) ID-v_{GS} plot of an NMOS transistor

$$g_m = \left.\frac{\partial I_D}{\partial v_{GS}}\right|_{v_{GS}=DCV_{GS}} = K'\frac{W}{L}\left(DCV_{GS} - v_{th}\right)\left(1 - \lambda v_{DS}\right) \qquad \text{Eq. 7.7}$$

Around DCV_{GS}, the current is I_{D1}, and consequently, Eq 7.6 and Eq 7.7 can be combined to express the current dependent transconductance:

$$\left. g_m \right|_{I_{D1}} = \frac{2I_{D1}}{\left(DCV_{GS} - v_{th}\right)} = \sqrt{2I_{D1}K'\frac{W}{L}(1 - \lambda v_{DS})} \qquad \text{Eq. 7.8}$$

From the high frequency small signal model of an NMOS transistor, as shown in Fig. 7.14, the ratio of I_{Source}over I_{Input} can be obtained to show the trends of the conversion gain of a transconductance stage.

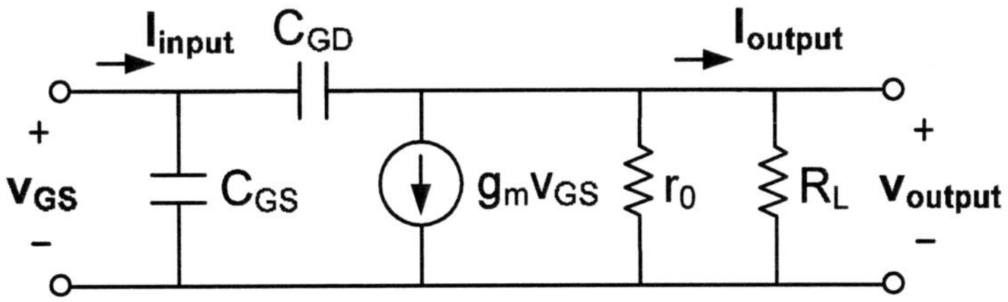

Fig. 7.14 High frequency small signal model of a NMOS transistor

First of all, if the current through C_{GD} is not taken into account in comparison to the current g_mv_{GS} (Sedra and Smith 2004), the output voltage v_{output} is:

$$v_{output} = -g_m v_{GS}\left(r_0 // R_L\right) \quad \text{Eq. 7.9}$$

And consequently, I_{output} is directly obtained by V_{output}/R_L:

$$I_{output} = \frac{-g_m v_{GS}\left(r_0 // R_L\right)}{R_L} \quad \text{Eq. 7.10}$$

For the input current, the Miller approximation may be used to obtained the following circuit, using $K= -g_m(r_0//R_L)$. C_{GD} is then separated into C_{GD1} and C_{GD2}:

$$C_{GD1} = C_{GD}\left(1-K\right)$$
$$C_{GD2} = C_{GD}\left(1-\frac{1}{K}\right) \approx C_{GD} \quad \text{Eq. 7.11}$$

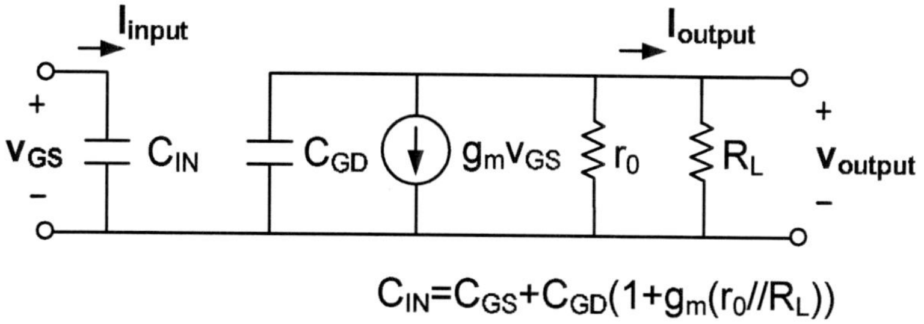

Fig. 7.15 High frequency small signal transformed model of a NMOS transistor

From Fig. 7.15, the input current is directly deduced as:

$$I_{input} = v_{GS} C_{IN}\, jw \quad \text{Eq. 7.12}$$

Eq 7.11 and Eq 7.12 can now be used in Eq.7.5 to obtain the conversion gain of a single transistor is proportional to the output and input current ratio:

$$ConversionGain \propto \frac{I_{Output}}{I_{Input}}$$

$$CG \propto \frac{\dfrac{-g_m v_{GS}\left(r_0 // R_L\right)}{R_L}}{v_{GS} C_{IN}\, jw} \quad \text{Eq. 7.13}$$

$$CG \propto \frac{g_m\left(r_0 // R_L\right)}{R_L C_{IN} 2\pi\ f}$$

After completing a simple inspection of the conversion gain of this stage, some first trend can be extracted for this subsection, but also for the frequency constraint design. The mixer gain is limited by the operating frequency of the RF signal as the conversion gain is conversely dependant on the frequency.

Another consideration to take into account is the size of the NMOS transistors for this transconductance stage. C_{IN} is composed by C_{GS} and C_{GD}, which are purely size-dependant, meaning that any increasing in the size of the transistors entails an increasing in the value of C_{IN} which here is clearly limiting.

Besides, the transconductance, g_m, is the only term which is current dependant and consequently the only parameter related tothe low power design constraint studied here. The expression of g_m in the strong inversion has already been set in Eq 7.8, but some authors suggest the use of the weak inversion or moderate inversion operating regions for this stage, and this is why the simplified expression of g_m in the subthreshold region is presented here:

$$g_m = \frac{I_D}{kT/q} \qquad \text{Eq. 7.14}$$

Nevertheless, the conversion gain of the switching transistor used in the transconductor stage is directly dependant of the current I_D.

Now, how is this single transistor study reflected in the conversion gains of the active mixers? It depends on their architecture. As seen in section 7.2, the Gilbert cell is the studied mixer circuit for low power, and so is the only one presented here.

7.3.1.1.2 Double Balanced Mixers: Gilbert Cell

This configuration, presented in subsection 7.2.1.1, is composed of a differential-pair driver stage (transconductance RF-driven stage) and a differential switching quad (LO-driven stage).

Apart from the fact that a differential-like architecture rejects the common mode noise and so the DC offsets at the baseband output are reduced, the double-balanced mixer has other advantages compared to its single-balanced counterpart; indeed, it eliminates the LO feedthrough, as well as both the interferers and noise superimposed to the LO signal applied to the mixer.

The upper stage is fed by the output current of the transconductance stage. Assuming basic simplifications for the sake of fast understanding, $g_m v_{RF}$ is the output current of this RF-driven stage. Consequently, the LO-driven stage output power is basically composed of $g_m v_{RF}$, a factor due to its switching characteristic, and R_L.

The differential structure avoids the odd harmonics from the mixing products of the LO large signal. When it is a square signal, the mathematical expression is presented in Eq 7.1. The first harmonic, which is the fundamental harmonic, has an amplitude of $A(4/\pi)$.

This leads to a ratio between the output voltage and the input voltage of

$$\frac{4}{\pi} g_m R_L \quad \text{Eq. 7.15}$$

7.3.1.1.3 Current Reuse Technique

The conversion gain is clearly dependent on the current. Then, the gain and the power consumption are directly related. In this field, there is a very effective technique to improve the relation between conversion gain and power consumption: the current reuse, also called charge injection. The Fig. 7.16 presents this alternative, adding the transistors M7 and M8.

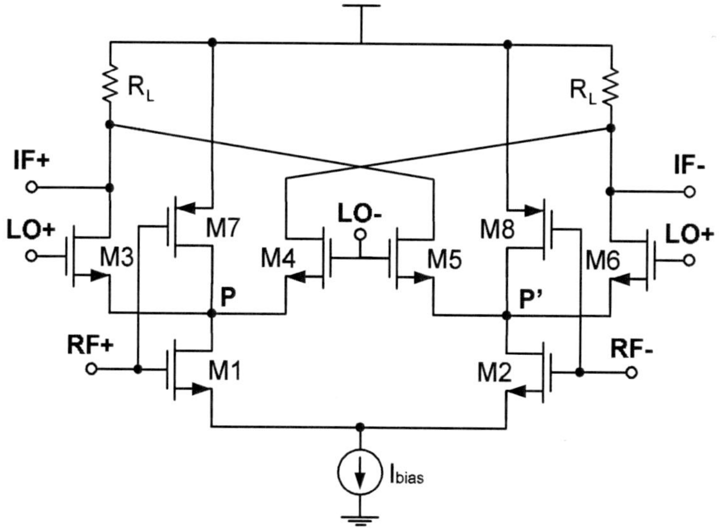

Fig. 7.16 Double balanced Gilbert cell active mixer with current reuse

Charge injection in active mixers has proven its viability with its contribution to the improvement of conversion gain, noise Fig. and switching efficiency (Darabi et al 2000). In order to save power consumption, this charge injection technique has been implemented in this mixer as part of the driver stage. Both transistors M7 and M8 inject charge into node P. This way, the injected current flows through the driver stage (M1 and M2) increasing its transconductance for a given bias current. In addition, since the bias current of the driver stage remains constant, the injected current makes the current through the switching FETs (M4-M6) lower, improving the switching efficiency. As a consequence and as it is analyzed in subsection 7.3.3.1, the low frequency noise is reduced. The optimum ratio between the current through the switches and the injected current is approximately 6%(Alvarado et al. 2008).

At the same time, the current injection transistors form part of the RF driving stage exploiting the current reuse technique (Karanicolas 1996) (section 5.1 of this book). With such a configuration, the overall transconductance of this stage is doubled (Eq 7.16) for the same bias current,

$$g_{mP} = g_{m1,2} + g_{m7,8} \qquad \text{Eq. 7.16}$$

where g_{mP} is the total transconductance at node P (Fig. 7.16), and both $g_{m1,2}$ and $g_{m7,8}$ are the transconductances of the NMOS and the PMOS devices of the RF stage (M1, M2 and M7, M8 respectively). A proper choice of transistor sizes and careful layout reduce the parasitic capacitance at node P minimizing the indirect mechanism contribution to flicker noise (subsection 7.3.3.1.1).

7.3.1.2 Gain in Passive Mixers

As shown in subsection 7.2.2 (Fig. 7.11), there is no supply voltage in these topologies, and consequently the input current cannot be positively amplified. Therefore, the conversion gain must be negative.

The mixing process is a multiplication of two signals which ideally provides an output signal at the sum and the subtraction of the input frequencies. Besides, as seen in section 7.1, in Eq 7.2, the fundamental component of a square wave is $4/\pi$ times the amplitude of the square wave. Then, assuming a multiplication of the RF input signal by a squared unit-amplitude LO signal, the value of the ideal conversion gain is (Lee 1998):

$$ConversionGain\Big|_{LOsquare} = \frac{2}{\pi} \qquad \text{Eq. 7.17}$$

However, with the correct DC polarization for the LO – driven switching transistors, there is a curious result if this conversion gain is compared with the one obtained with a sinusoidal LO signal.

(Lee 1998) explains in his book that the overall conversion gain is greater with a sinusoidal drive because of the $\pi/2$ ratio from peak to average of the signal applied. The resulting conversion gain in this case is:

$$ConversionGain\Big|_{LO\sin e} = \frac{\pi}{4} \qquad \text{Eq. 7.18}$$

Moreover, the conversion gain for the LO square signal may differ from the value shown in Eq 7.17 from the power consumption point of view. The mixing of the IF signal with a LO square signal directly affects the power consumption of the buffer of the LO signal, which then largely increases.

As stated in (Lee 1998), the power consumption of the LO drivers may be potentially reduced by a factor of Q^2, if the gate capacitance of the switches is resonated with an inductor. This variation has the drawback of the bandwidth limitation provided by this LC parallel structure. It is then appropriate for narrow-band applications. But in return, it can result in a benefit because of the inner filtering applied to the input signal, avoiding undesired saturation of the transistor switching.

7.3.2 *Linearity*

The first concern about the linearity of CMOS mixers (double or single active balanced and passive ring mixers) is the switching of the transistors which is highly dependant on the LO signal. If the LO-driven transistors act as ideal switches, the linearity is not affected by this operation. This evidence leads to 2 conclusions from the linearity point of view:

- The passive mixers have a better linearity characteristic.
- The active mixers are mainly limited by the transconductance stage.

In any case, the studies on conversion gain based on small-signal approximations, only hold at a reduced input power. The gain compression and the intermodulation distortion are the main problems if the input signal is increased.

7.3.2.1 Linearity in Active Mixers

As seen in the introduction of this subsection, the linearity of the active mixer is principally dependant on the transconductance stage. The following subsections analyze several issues regarding linearity constraints in active mixers: the polarization region of the transconductance stage, source degeneration and other linearization techniques.

7.3.2.1.1 Polarization Region and Linearity

Considering the work of Hsieh and Lu in (Hsien 2007) on ultra low voltage RF front-ends, the drain current of the transconductor stage can be expressed as a third-order power series, as follows:

$$I_D = c_1 v_{GS} + c_2 v_{GS}^2 + c_3 v_{GS}^3 \qquad \text{Eq. 7.19}$$

where c_i are the coefficients of the i-order harmonics of this current. The major contributor to the distortion is c_3, and an approximation of the IIP3 characteristic is:

$$IIP_3 \approx \frac{2}{3Z_{IN}}\left(\frac{c_1}{c_3}\right) \qquad \text{Eq. 7.20}$$

where c_1 is the transconductance of this stage's NMOS transistor (g_m), and Z_{IN} its input impedance. As in the case of the conversion gain, the power consumption constraint is then clearly dependant on g_m.

This work has proposed a current reuse architecture based on the Gilbert cell for the downconversion mixer, studying in depth the evolution of the coefficients c_1 and c_3 and their relation with v_{GS}. The evolution of the v_{GS}-c_3 plot (Fig. 14 (a) and (b)) of (Hsien07) shows a zero crossing from the positive plane to the negative one, and this specific point is considered as a "sweet spot" for the linearity. No current is engaged in the c_3 definition, but the improvement of the IIP_3 with the same current consumption is a design constraint from the linearity point of view.

7.3.2.1.2 Source Degeneration and Linearity

The linearity of the transconductance stage can be improved using a degeneration impedance. This helps in the linearization of the transfer characteristic, which is evenmost effective if the admittance looking from the source terminal of the transistor is much larger than the conductance of R_{SOURCE} (Lee 1998).

Fig. 7.17 shows the small-signal model of a degenerated amplifying transistor.

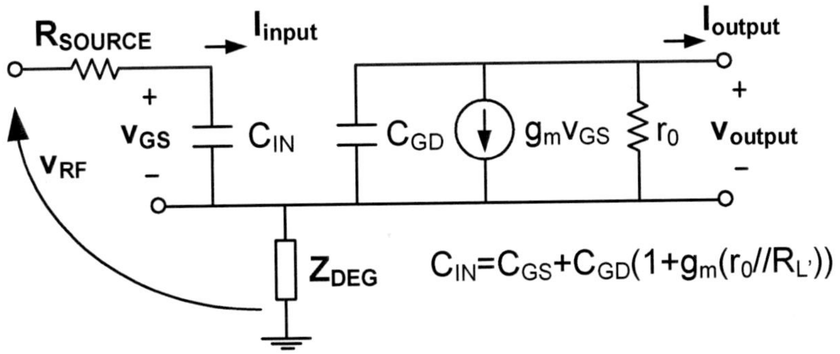

Fig. 7.17 NMOS transistor with degeneration impedance small signal model

where $R_{L'}$ is the impedance seen from the source of the transconductance stage. With this small signal model, the IIP_3 is affected by the change of the input impedance Z_{IN}. From Fig. 7.17, the following expression for Z_{IN} can be deduced:

$$Z_{IN} = \frac{R_{SOURCE} + C_{IN} jw + Z_{DEG}(C_{IN} jw + g_m)}{C_{IN} jw} \qquad \text{Eq. 7.21}$$

According to Eq 7.21, if Z_{DEG} is an inductor with a value L_{DEG}, the third addend is:

$$L_{DEG} jw(C_{IN} jw + g_m) = -L_{DEG} C_{IN} w^2 + L_{DEG} g_m jw \qquad \text{Eq. 7.22}$$

The first resultant term is negative, which means that the combination of the inductor and the capacitor produces a lowering of the Z_{IN}. Consequently, this is beneficial for the IIP_3, without having to change the current consumption of the basic amplifying stage.

The conclusion is that the inductive degeneration is a design technique extremely useful from the low power consumption point of view, but it trades off the conversion gain of the block.

A similar reasoning may be adopted to include an inductor in the RF input connection. It is then connected in series with the R_{SOURCE} and the C_{IN} capacitor. It then also helps in decreasing the Z_{IN} because of a combined negative addend, which improves the IIP_3. Some studies have used this technique successfully

(Lee 1998), but it has to be taken into account that the effective bandwidth at the input is highly limited by the LC tank that appears in series with the RF source. The value of this added inductor has to be carefully chosen to match the oscillation of the LC structure.

7.3.2.1.3 Other Linearization Techniques

Other more complex techniques, using feedback structures have been proposed by several authors for double balanced mixers (Lee 1998).

The most common alternative found in the state of the art to push active mixers to higher linear behavior is the multi-tanh arrangement(Gilbert 1998). This is based on the evidence that the transistors are inherently linear in a small range over the correct polarization. It is then wise to share the linear responsibility over several transconductance cells to obtain an extended linearity, although the current consumption is also increased by the factor of the number of g_m cells, which is not desirable from the power consumption point of view, it nor is recommended from the power constraints point of view.

7.3.2.2 Linearity in Passive Mixers

In this subsection, no nonlinear effect has been taken into consideration for the LO-driven transistors, as the CMOS technology presents an excellent switching behavior with the appropriate v_{GS}, which confers to the CMOS passive mixers a higher linearity characteristic. The RF signal here is directly mixed with the LO signal in the voltage domain.

Nevertheless, the negative conversion gain of this topology is a drawback also from the linearity point of view. Not as a single circuit, but as a circuit embedded in a receiver. The system overall linearity is related with the system overall gain, and consequently in the case of using a passive mixer, the following block should perform a better linearity characteristic than it would do with an active mixer. This also has to be taken into account for the power constraint of a system. Similar trade-off will be considered for the noise performance, as stated in the next subsection.

7.3.3 Noise

This subsection deals with the mechanisms that contribute noise in the Gilbert cell based mixers. Due to the signal frequency translation, the low frequency noise plays an important role in mixer performance, and is analyzed in the next subsection.

7.3.3.1 Low-Frequency Noise Analysis

Two different mechanisms are presented in the Gilbert-cell based active mixer architectures leading to low frequency (1/f) noise: the direct mechanisms and the indirect mechanisms. They are outlined in the following paragraphs.

7.3.3.1.1 Direct Mechanisms of Low Frequency Noise

The main source of flicker noise in a Gilbert cell based mixer is the switching FETs. These transistors experience a random fluctuation of charge, leading to a "direct mechanism" of flicker noise (Darabi et al. 2000). The time constant of this flicker charge fluctuation is very low compared to the RF frequency. As a consequence, a noise voltage (V_n) appears at the gate of one of these transistors with a spectral density proportional to 1/f, modulating the transistors' switching time and therefore modifying the zero-crossing instants by $\Delta t = V_n(t)/S$, where S is the slope of the LO voltage at switching time. This noise leads to a current at the mixer output whose average value over one period is given by Eq 7.23:

$$i_{out,noise} = 4I\frac{V_n}{ST} \qquad \text{Eq. 7.23}$$

where I is the amplitude of the output current and T is the period of the LO. The low frequency noise voltage (V_n) appears at the output without frequency translation and degrades the baseband signal. The SNR at the mixer output can be used to estimate its noise figure

$$SNR_{out,direct} = \frac{ST}{2\pi(V_{GS} - V_t)}\frac{V_{in}}{V_n} = \frac{2A}{(V_{GS} - V_t)}\frac{V_{in}}{V_n} \qquad \text{Eq. 7.24}$$

where the last term represents the SNR for a sinewave LO of amplitude A, and (V_{GS} - V_t) is the transconductance FETs overdrive voltage. On the other hand, the flicker noise of a MOS transistor is given by the empirical formula (Liu 1998)

$$v^2_{flicker,MOS} = \frac{k_f I_{ds}^{af}}{C_{ox} L_{eff}^2 f^{ef}} \qquad \text{Eq. 7.25}$$

where C_{ox} is the unit capacitance of oxide, k_f is a device-specific constant, L_{eff} is the effective channel length of the MOS, and I_{ds} is the channel bias current. *af* and *ef* are current and frequency index respectively.

Looking at Eq 7.24, the noisefigure can be lowered by means of:

- Increasing the ST product
- Increasing the switching FETs gate area in order to reduce the noise voltage (Eq 7.23)
- Diminishing the overdrive voltage, taking into account that the two latter issues represent a trade-off with bandwidth.

Considering that the LO frequency is given by the application and cannot be changed, several approaches have been commonly used to reduce the flicker noise in a MOS transistor:

- Use longer channel devices in the flicker noise sensitive areas, such as the tail current source or in the switching transistors, but degrading transconductance and switching time

- Decrease k_f using PMOS devices, but leading to similar consequences as in the previous approach due to their low mobility
- Lower the transistor bias current, compromising the transconductance and linearity.

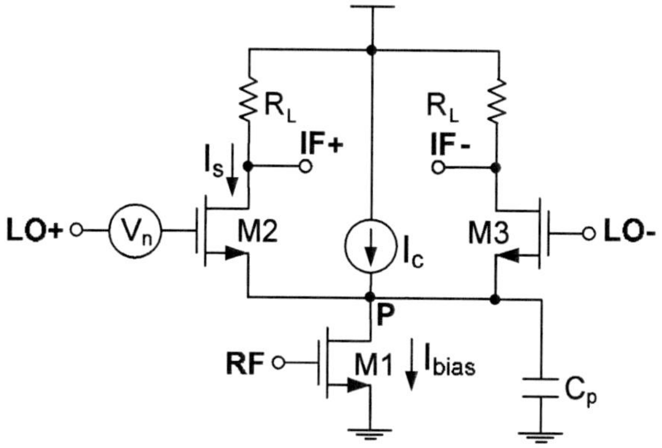

Fig. 7.18 A single-balanced simplified mixer with charge injection

In a mixer, the charge injection technique (Goo-Young 2007), (Lee et al. 2000) allows reducing the current in the switching FETs maintaining the same current in the input stage and therefore not compromising transconductance and linearity. In the single-balanced mixer shown in Fig. 7.18I_{bias} can be kept at its proper value while I_s can be reduced by injecting more current, I_c, at node P. This technique has been implemented in one of the examples illustrated in the examples subsection, as presented in 7.4.1.

7.3.3.1.2 Indirect Mechanisms of Low Frequency Noise

From the previously outlined analysis, it can be concluded that if a perfect square wave signal is applied as LO (constant amplitude, duty cycle and infinite slope in the zero-crossings) the flicker noise will be suppressed (Eq 7.23). However, this is not totally true, as flicker noise is still present by what is called "indirect mechanism". This noise depends on the LO frequency and the parasitic capacitance at the tail of the LO differential pair, C_p. The 1/f noise spectrum at the mixer output is given by Eq 7.26:

$$i_{out,noise}(f) = 2\frac{C_p}{T}V_n(f)\frac{(C_p\omega_{LO})^2}{g_{ms}^2 + (C_p\omega_{LO})^2} \qquad \text{Eq. 7.26}$$

where g_{ms} is the transconductance of the switching FETs. It can be inferred from Eq 7.26 that the flicker noise can be reduced by decreasing the parasitic capacitance at the tail C_p or by decreasing the LO frequency, which is not always possible as it is application dependant. When, $\omega_{LO} \ll g_{ms}/C_p$ in a sine curve LO the contribution of this mechanism to low frequency noise is negligible. It can be concluded that in an active double-balanced mixer only the direct mechanisms in the switching stage contribute to the flicker noise.

7.3.3.2 High Frequency Noise

With regard to high frequency noise, all the stages of the mixer contribute to overall white noise (Terrovitis et al 1999), the total noise voltage at the output of the mixer is shown as

$$\hat{v}^2_{out,noise} = 8kTR_L\left(1 + \gamma\frac{2R_L I_s}{\pi A} + \gamma g_m R_L\right) \qquad \text{Eq. 7.27}$$

where γ is the channel noise factor, T is the absolute temperature, k is the Boltzmann constant, A is the amplitude of the LO and g_m is the transconductance of the RF stage.

The first term of Eq 7.27 is due to the load resistors, the second refers to the contribution of the switches, and the final term originates from the noise introduced by the transconductance stage and transferred through the mixer output. It can be deduced that a low DC current in the switching transistors, as well as a relatively high LO amplitude help reducethe noise figure of the mixer.The former is evident, and the latter, as discussed before, has wider implications on the noise issues asthis amplitude requires higher power consumption in the LO signal drivers.

Linearity and gain are here also compromised and a trade-off has to be established for each case studied. A modest LO swing isneeded to keep the transistors in the saturation region and then bearon the linearity characteristic, but this negatively affects the noise.

The degeneration is primarily used to improve the linearity of the mixer but, as pointed out in (Fong 1999) the reactive degeneration (inductive or capacitive) results in lower NF than that with resistive degeneration. In fact, the added inductance does notadd thermal noise and consequently, the noise figure degradation is lower.

7.3.4 Bandwidth

The bandwidth relation with the power consumption follows a classical pattern. For a constant conversion gain, NF andIIP$_3$, higher frequency may be reached if a higher current is applied.

The first is the easiest to explain. By inspection of Eq 7.13, the conversion gain is directly dependant on the output to input current ratio. This equation shows that the frequency is on the denominator whereas the transconductance is in the numerator. That means that for the sake of reducing the power consumption of the mixer, the RF input frequency should be decreased. Unfortunately, this is not a configurable parameter from any system, as it is one of the specifications. Consequently, to attain the desirable gain, a higher current has to be applied, and then a higher power has to be consumed. High frequency and low noise are also inversely related as shown in Eq 7.25.

The linearity trade-off with the current is slightly different. Initially, the IIP3 expressions are not clearly related to the frequency, but to the bandwidth of the input signals. There are some linearity improvement techniques, such as the inclusion of an inductor in series with the RF signal. Indeed, it always has to be included as the bondwire's simplest model is an inductor. As commented in the subsection dedicated to linearity, the limitation for the frequency is not an upper limit, but a restricted bandwidth, which depends on the added $L_{GATE}$and C_{IN} values.

7.3.5 *Impedance Matching and Port Isolation Considerations*

As previously mentioned in subsection 0, impedance matching and port isolation are directly related to the transmission and reflection of the signals (RF, IF and LO) through the different ports.

7.3.5.1 Impedance Matching Considerations

When a port impedance is not perfectly matched to that of the source resistance, some of the power delivered to the port is then reflected back to the source. Any mismatch is a waste of power as is a waste of signal. Considering the inputs of the mixer, the RF signal, as well as the LO signal, should be perfectly matched to their input circuits,if not reflection of energy occurs.

The most seriousproblem is the lossof the RF signal as itis the weakest one, thus highly seriously affecting the mixing performance. Nevertheless, the reflection of the high-powered LO signal also causesproblems to the system as the power reflected leaks through the LNA reverse path and isre-radiated. Both kind of mismatching have to be avoided.

For both, special care has to be taken of the gates' inductor added by the final system packaging (bondwires). A careful analysis of the RF and LO input nets has to be performed to assure the best matching. The power consumption here is also implied in the trade-off.

Furthermore, as explained in subsection 7.3.1, the LO power is also dependant on the power consumption dedicated to the drivers of this signal.

7.3.5.2 Port Isolation Considerations

The isolation of the three ports of a mixer is key to the overall power consumption of the RF front-end. Low-IF, or even better Zero-IF, architectures are undoubtedly well suited for lowering the power consumption of the whole system.

In this case, as presented in 0, the isolation between the RF and the LO ports becomes critical as the two injected signals are quite close (even superimposed for Zero-IF) in the frequency domain.

Two paths should be taken into account (C_{GS} and C_{GD}) in a Gilbert cell topology, but the differential operation in terms of LO signal cancels out the common mode nodes. In fact, the RF - LO isolation depends solely on the mismatch between the MOS devices and the loads. There is no power constraint at this point; only a careful layout is key to the port isolation.

7.4 Low-Power Mixer Design Examples

This section describes in detail the design of three different low-power mixer examples, regarding its circuit design considerations, layout implementation, the test setup used for the measurement of their main parameters, the characterization results, and finally a short discussion on such results.

- The first example is a low-power and low-noise Gilbert cell based active mixer for the DVB-T and DVB-H standards (DVB-T_EN300744), (DVB-H_EN302304).
- The second example shows another low-power Gilbert cell based active mixer for higher frequency applications in the 5GHz U-NII frequency band, such as WLAN (IEEE99), (IEEE03).
- Finally, the third example analyzes a passive mixer, again for the 5GHz U-NII frequency band.

7.4.1 Example 1: Low-Power Low-Noise Mixer for DVB-T/H

This subsection includes the description of a low-power and low-noise mixer for DVB-T/H. It comprises circuit design issues, layout implementation considerations, the description of the test setup for the measurement of its main parameters, the characterization results as well as a brief discussion on such results with regard to the stateoftheart.

7.4.1.1 Circuit Design

With the above considerations (section 7.3), a double-balanced CMOS direct conversion mixer has been designed to be compliant with both DVB-T and DVB-H standards. Its simplified schematic is depicted in Fig. 7.19.

Apart from the fact that a differential-like architecture rejects the common mode noise and so the DC offsets at the baseband output are reduced, the double-balanced mixer has other advantages compared to its single-balanced counterpart; indeed, it eliminates the LO feedthrough, as well as both the interferers and noise superimposed onthe LO signal applied to the mixer.

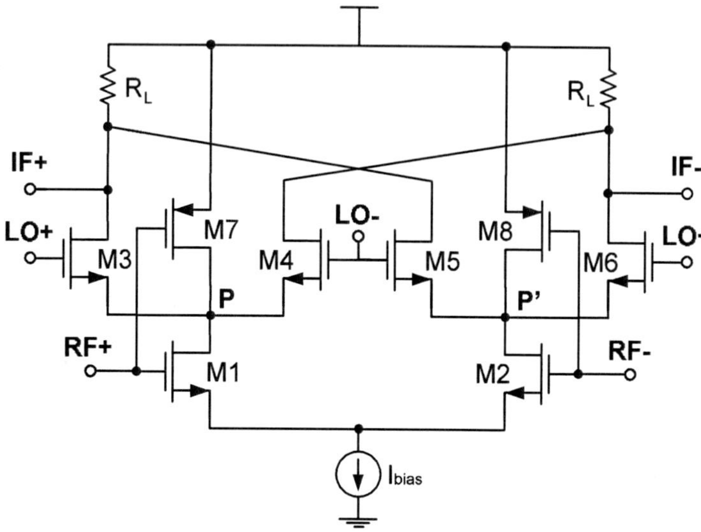

Fig. 7.19 Zero-IF mixer for DVB-T/H simplified schematic

Charge injection in active mixers has proven its viability with its contribution to the improvement of conversion gain, noise figure and switching efficiency (Darabi et al. 2000). In order to save power consumption, this charge injection technique has been implementedin this mixeras part of the driver stage. Both transistors M7 and M8 inject charge into node P. This way, the injected current flows through the driver stage (M1 and M2) increasing its transconductance for a given bias current. In addition, since the bias current of the driver stage remains constant, the injected current makes the current through the switching FETs (M4-M6) lower, improving the switching efficiency. As a consequence, the low frequency noise is reduced (Eq 7.24). The optimum ratio between the current through the switches and the injected current is approximately 6%(Alvarado et al. 2008).

At the same time, the current injection transistors form part of the RF driving stage exploiting the current reuse technique (see 7.3.1.1.3). With such a configuration, the overall transconductance of this stage is doubled (Eq 7.16) for the same bias current. A proper choice of transistor sizes and careful layout reduce the parasitic capacitance at node P, minimizing the indirect mechanism contribution to flicker noise (Eq 7.26). The width of the NMOS pair has been set at 150μm, while the width of the charge injection pair is four times lower. With the exception of tail transistors, all channel lengths have been set to the minimum allowed by the technology, which is, 0.35μm.

Polysilicon resistor loads have been chosen due to their high bandwidth. In addition, they are free of flicker noise, which is a key point in such zero-IF devices. Since they contribute white noise according to (Eq 7.27) their value must be carefully selected to guarantee an appropriate conversion gain with no NF penalty. Using the charge injection technique, lower current flows through the switching transistors and hence through the loads, reducing the voltage headroom problem. Therefore, higher resistance values can be used to increase the conversion gain. For this design, a value of 3kΩ has been selected for both resistors. For on-wafer measuring purposes, a differential source follower buffer has been added to the output in order to drive the 50Ω impedance of the testing equipment.

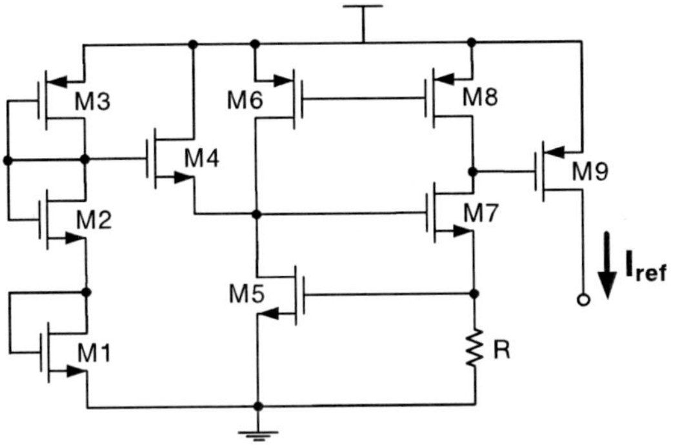

Fig. 7.20 Voltage compensated current source

Apart from the voltage supply, another important aspect for the polarization of the different circuit elements is the bias current. In order to guarantee stable operating points against random variations in the power supply, a voltage compensated current source has been implemented (Fig. 7.20). This current source provides the mixer core with a stable current of 2mA for variations in the voltage supply up to a 30%. The bias current of the output buffer is also 2mA, and is used only for measurement purposes.

7.4.1.2 Layout Implementation

The layout of the mixer core is shown in Fig. 7.21. In order to guarantee a good balance between the differential branches of the circuit, special care has been taken regarding the component symmetry. In addition, other matching techniques such as the common-centroid (Hastings 2000), (Clein 2000) have been implemented, in order to reduce the component dispersions through the silicon wafer.

Fig. 7.21 Layout of the DVB-T/H mixer core

With regard to noise behaviour optimization, it is important to place as many contacts as possible from the substrate to ground, in order to increase the isolation between components and to reduce the noise coupling through the substrate. Guard-rings have been also implemented in all elements of the layout, as well as substrate contacts in the vicinity of the FETs' gates of the transconductance stage, reducing their contribution to white noise.

In order to de-couple the noise coming from the power supply, as much capacitance as possible has been implemented from V_{DD} to ground (40pF approx.), by means of:

- Tracking wide V_{DD} paths. This way, the parasitic capacitance from the track to the substrate directly adds to the V_{DD}-to-ground capacitance.
- Laying out as many capacitors as possible from the V_{DD} paths to ground. For the on-wafer measurement of the mixer, the free space left within the pads and the mixer core has been filled by capacitors.

The complete layout of the mixer is shown in Fig. 7.22. An inner guard-ring around the mixer core and an outer guard-ring enclosing the pads have been implemented for improved system isolation and shielding.

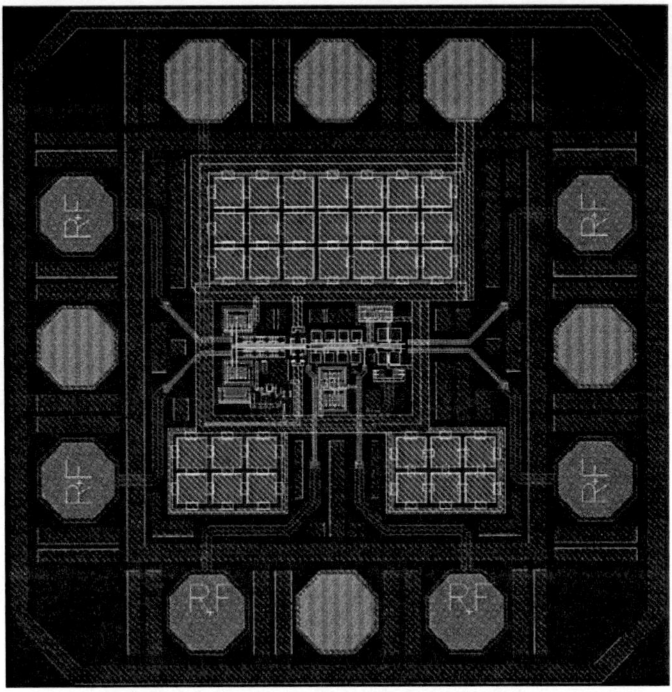

Fig. 7.22 DVB-T/H mixer complete layout

7.4.1.3 Mixer Test Setup and Characterization Results

The mixer has been measured on-wafer. The die photograph is shown in Fig.7.23. A detailed view of the mixer core can be observed in Fig.7.24.

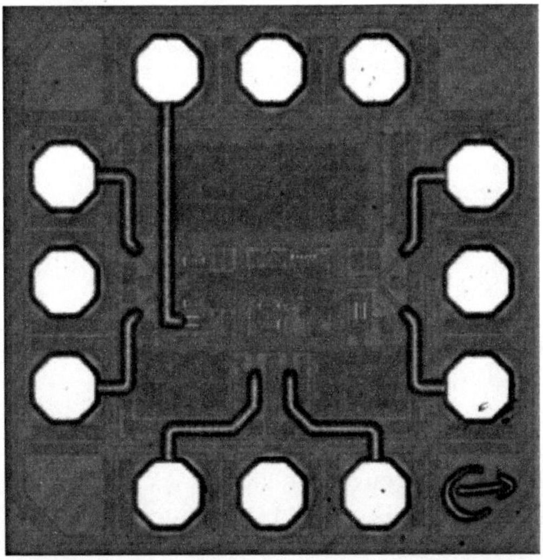

Fig. 7.23 DVB-T/H mixer die photograph

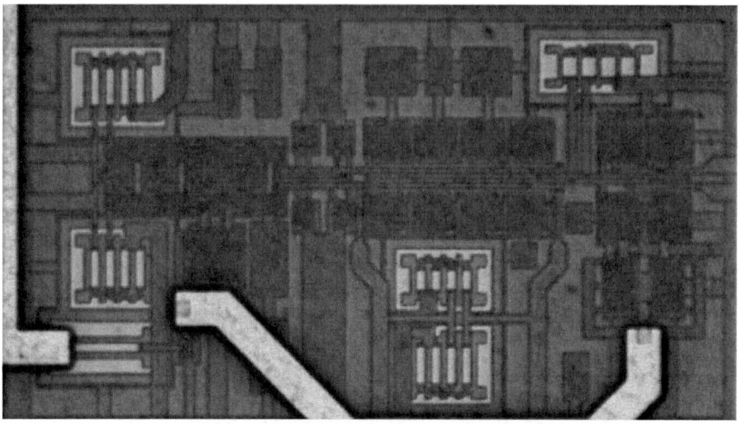

Fig. 7.24 DVB-T/H mixer core die photographe

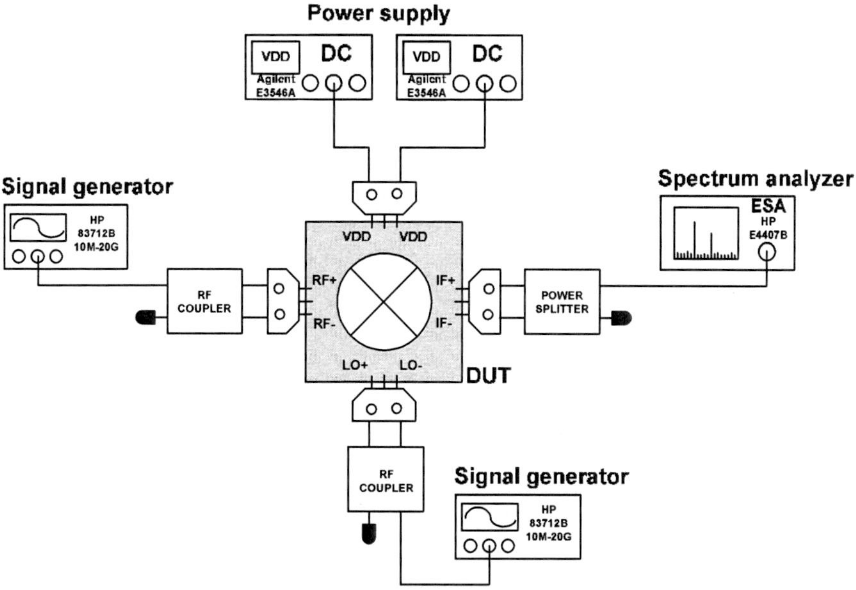

Fig. 7.25 Mixer conversion gain measuring setup

Fig.7.25 and Fig.7.26 show the test setup for conversion gain and noise figure measurement respectively.

Plotted in Fig. 7.27 are the conversion gain and the noise figure measured at 4MHz IF, for an RF input frequency range extending from 100MHz to 2.5GHz. For the DVB-T/H band, the conversion gain varies from 19.5dB to 20.4dB, while the noise figure (DSB) is nearly constant (from 7dB to 7.2dB).

Fig. 7.28 shows the measured conversion gain and noise figure against the LO power. The flicker noise results have been extracted from post-layout simulations, due to the fact that the lowest frequency of the ESA is 300kHz. They represent the flicker noise (as the noise figure at 10kHz) against the LO power for different ratios between the injected current and the current through the switches (see 7.3.1.1.3 and 7.4.1.1).

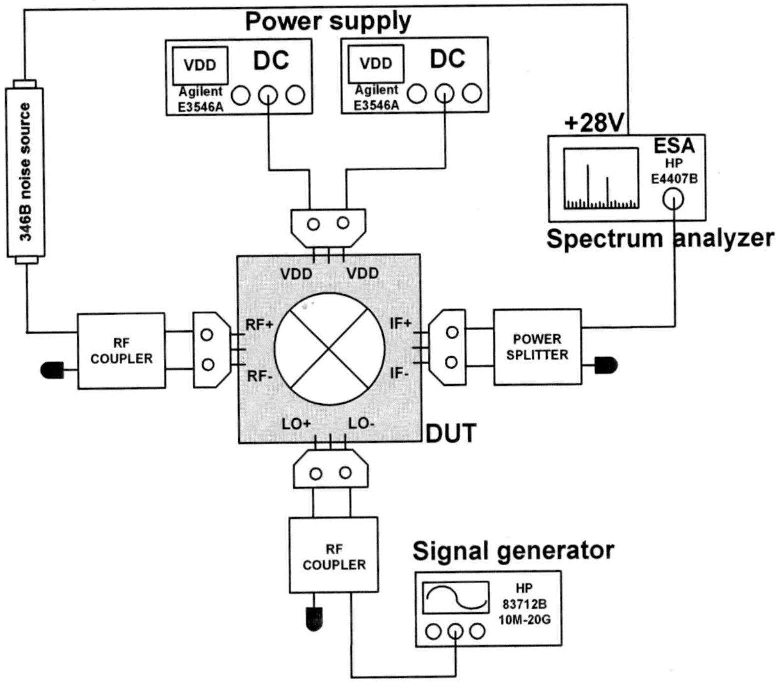

Fig. 7.26 Mixer noise figure measuring setup

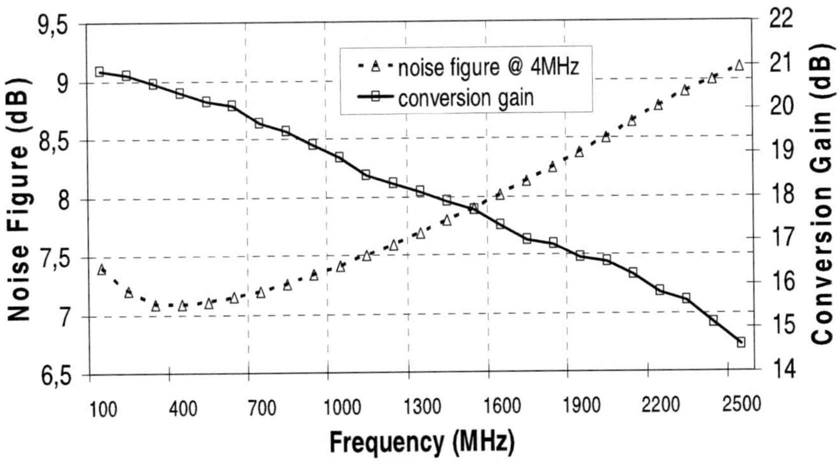

Fig. 7.27 DVB-T/H mixer conversion gain and NF measured results vs. RF input frequency (MHz)

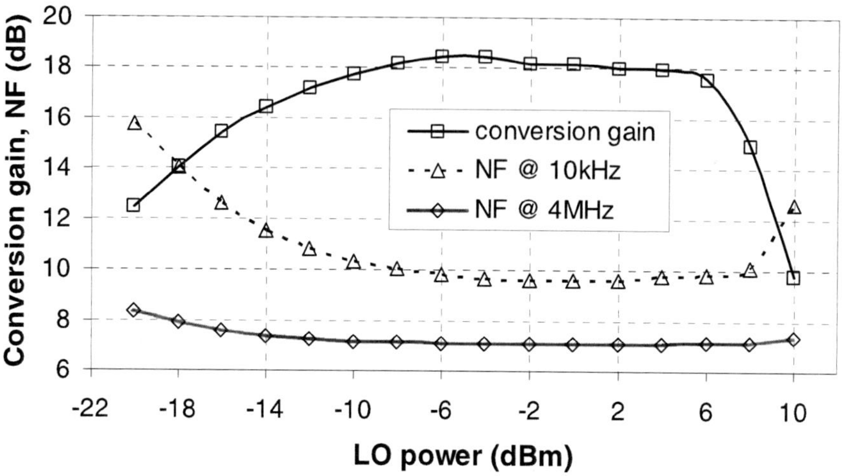

Fig. 7.28 DVB-T/H mixer conversion gain and NF vs. LO power (dBm)

On the other hand, Fig. 7.29 shows the measuring setup for the two-tone test. The measured mixer IIP3 is -8dBm, measured at 860MHz, as depicted in Fig.7.30.

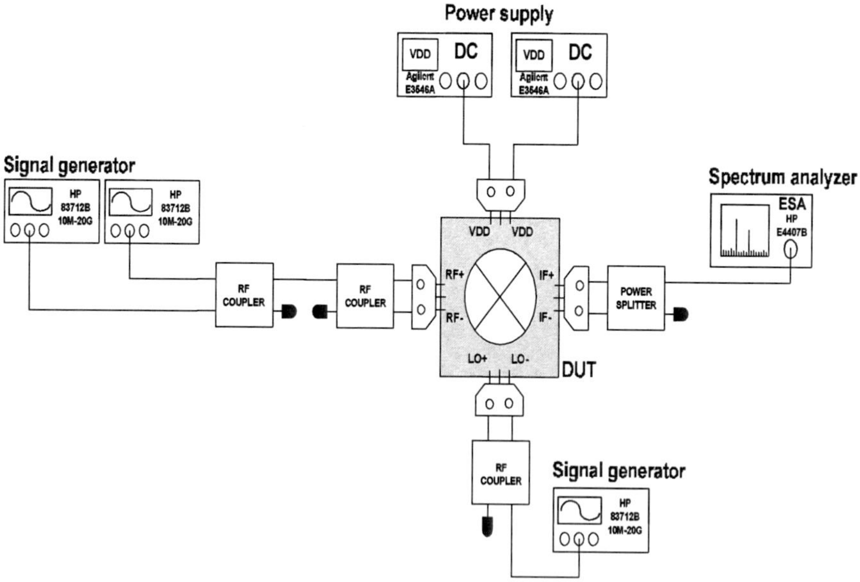

Fig. 7.29 Mixer two-tone test measuring setup

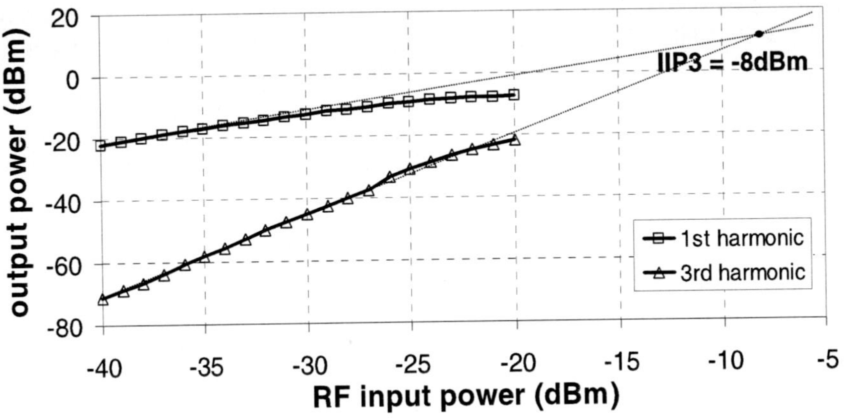

Fig. 7.30 DVB-T/H mixer third order intercept point (IIP3)

Finally, Table 7.1 summarizes all the mixer measured results and their initial specification. The power consumed by the output buffer (2mA) is excluded from the results summary as it is implemented just for on-wafer measurement purposes.

Table 7.1 DVB-T/H mixer measurements results summary

Parameter	Unit	Specification	Measurement
Conversion gain	dB	12	20
NF	dB	10	7.1
IIP3	dBm	0	-8
Current consumption	mA	min	2.3
LO power	dBm	min	-10

7.4.1.4 Discussion

With a current consumption of 2.3mA, the mixer performance has been characterized as outlined in the previous subsection. The most relevant parameters are highlighted below:

- The conversion gain is more than 20dB for the DVB-T/H frequency band
- The maximum noise figure (measured at 4MHz IF) is 7.1dB (DSB)
- The input referred third order intercept point (IIP3) is -8dBm
- The required LO power is -10dBm

Depicted in Fig. 7.27 are the frequency responses of the conversion gain and noise figure respectively. Although both parameters have been characterized up to 2.5GHz, a variation of 1dB (conversion gain) and 0.2 dB (NF) have been achieved for the DVB-T/H frequency band of interest. The noise figure has been characterized at an IF frequency of 4MHz (Double Side Band) since it is the upper bound of the DVB-T/H baseband signal. Therefore, the effective noise figure for the zero-IF front-end is 3dB lower, that is, 4.1dB, with an LO power of -10dBm.

Similarly, Fig. 7.28 shows the conversion gain and the noise figure as a function of the LO power. The response of both parameters is flat from an LO power of -12dBm up to +7dBm. For the reasons stated in sections 7.3.1 and 7.3.5 an LO power of -10dBm has been selected for the mixer performance. Likewise, depicted in Fig. 7.28 are different plots of the noise figure at an IF frequency of 10kHz (flicker noise). These plots represent different flicker noise responses depending on the I_s/I_c ratio (see 7.4.1.1). For the selected LO power, the flicker noise decreases up to an optimum value(I_s/I_c= 6%); this value has been selected for the mixer implementation, which leads to a noise figure of 11dB at an IF frequency of 10kHz, reducing the effects of the flicker noise in the zero-IF front-end implementation (section 3.6).

Finally, the linearity has been characterized by means of the input third order intercept point (IIP3). As depicted in Fig.7.30, the IIP3 does not fulfil the specification (IIP3 = -8dBm < 0dBm). However, post-layout simulations have predicted an increase in IIP3 of 10dB with no buffer connected to the output. Hence, the mixer core will present an IIP3 of +2dBm (within the whole front-end implementation), due to the fact that the output buffer has been implemented only for measurement purposes.

Shown in Table 1-2 is a collection of the most representative active mixers of the stateoftheart. These mixers are compared with the one implemented in this work (section 7.4.1) in terms of its most relevant parameters:

- Frequency of operation
- Conversion gain
- Double-sideband noise figure
- Input referred third order intercept point (IIP3)
- Power consumption
- Fabrication technology

It is difficult to state a performance comparison between the mixerpresented in this section with other devices in the state of the art, due to the lack of specific devices for DVB-T/H. However, such a comparison has been carried out with mixers working at similar frequencies.

Regarding conversion gain, the mixer presented in this work shows less than 1dB gain drop with respect to the values exhibited by (Phan et al 2005) and (Zencir et al. 2002). In addition, the results of (Phan et al. 2005)were extracted from simulation and (Zencir et al. 2002)showedworse noise figure and linearity. The NF reported in this example is the lowest (4.1dB SSB). With regard to linearity, as outlined in subsection 7.4.1.1, this mixer presents an IIP3 of +2dBm with no output buffer, only surpassed by (Phan et al. 2005) and (Shaikh and DeGroat 2005), at the cost of a poorer performance concerning gain, NF and power consumption.

Table 7.2 State-of-the-art active mixers for DVB-T/H frequency band

(REF)	Frequency (GHz)	Conversion gain (dB)	DSB NF (dB)	IIP3 (dBm)	Power consumptio n (mW)	Technology process gate length(μm)
(Ming-Feng et al 2004)	0.9	9.17	---	-5	1.35	0.18 CMOS
(Goldfarb et al. 2003)	2.16	18	8	-6.6	12.6	0.35 SiGe BiCMOS
(Phan et al. 2005)	1.32	20.5	5.6 (SSB)	+11.3	11.5	0.18 CMOS
(Zencir et al. 2002)	0.435	20.8	9.2	-13.4	5.4	0.5 CMOS
(Xuezhen et al. 2003)	5.8	7	14.3 (SSB)	-3	6.89	0.18 CMOS
(Bautista et al. 2000)	0.815	14.5	12	+2.4	10.8	0.35 BiCMOS
(Shaikh and DeGroat 2006)	2.1	16	13.8	+12.1	9	0.18 CMOS
(Sang-Sun et al. 2005)	5.15-5.825	10	12.3	+1.7	8.8	0.18 CMOS
(Alvarado et al 2008)	**0-900**	**20**	**7.1**	**-8**	**6.4**	**0.35 CMOS**

7.4.2 Example 2: Low-Power Mixer for WLAN (5GHz U-NII Band)

This subsection includes the description of a low-power mixer for WLAN (in the 5GHz U-NII frequency band). It comprises circuit design issues, layout implementation considerations, the description of the test setup for the measurement of its main parameters, the characterization results as well as a brief discussion about such results with regard to the stateoftheart.

7.4.2.1 Circuit Design

Fig. 7.31 depicts the double balanced core configuration selected for the WLAN (5GHz U-NII band) active mixer. As in the DVB-T/H mixer example, a balanced topology has been adopted to prevent the LO noisy products from disturbing the mixer output.

Concerning the power consumption, the most straightforward way to reduce the current needed by the device is by lowering the transistors' sizes. Besides, the operation in the 5GHz U-NII band recommends a minimum length for both sets of transistors. Unfortunately, the minimization of the width dimension is not possible as the noise and the linearity deteriorate due to the current decreasing to its minimum. The decisions about the transitions sizes require trade-offs.

The RF stage transistors have been extracted from the NMOS RF optimized devicesof the UMC CMOS 0.18μm 1P6M technology. TheNMOS transistors have the minimum length and a 30μm width shared among 5 fingers for a compacted layout. On the other hand, the LO stage has also been implemented with the minimum length allowed by the technology, and its transistors' width is 10μm.

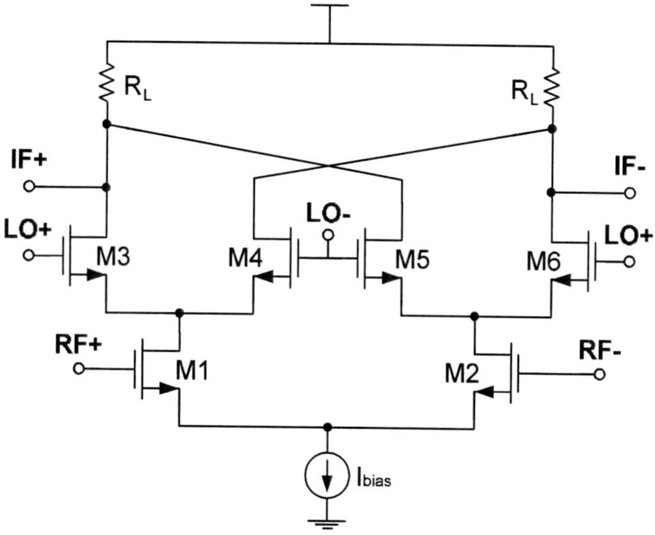

Fig. 7.31 Active double balanced mixer core configuration

Low dimensions for the transistors of the active mixer have reduced the power consumption of this device. Besides, the gain requirement has not been as restrictive as the NF, which benefited from the low distortion caused by the 1.9mA of this core. The resistive load is 1kΩ. The buffer is also a common source amplifier and consumes the remaining specified current up to the current consumption of 4.9mA.

7.4.2.2 Layout Implementation

The layout issues regarding the implementation of the mixer are discussed in subsection 7.4.1.2. Depicted in Fig. 7.1 is the layout of the mixer core. The whole view of the mixer layout is shown in Fig. 7.33.

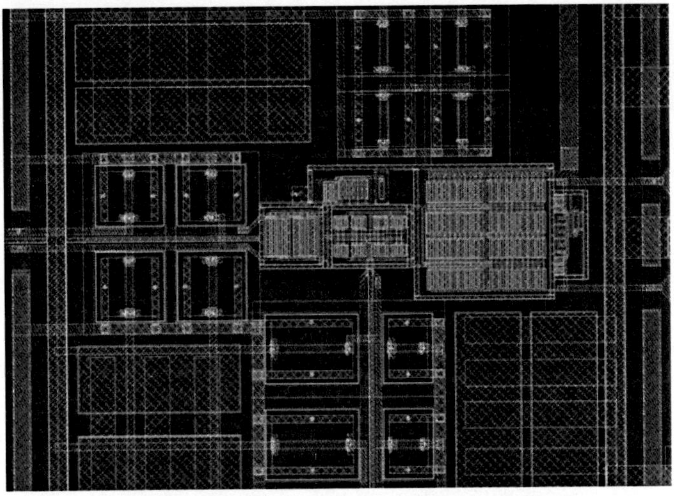

Fig. 7.32 Layout of the WLAN active mixer core

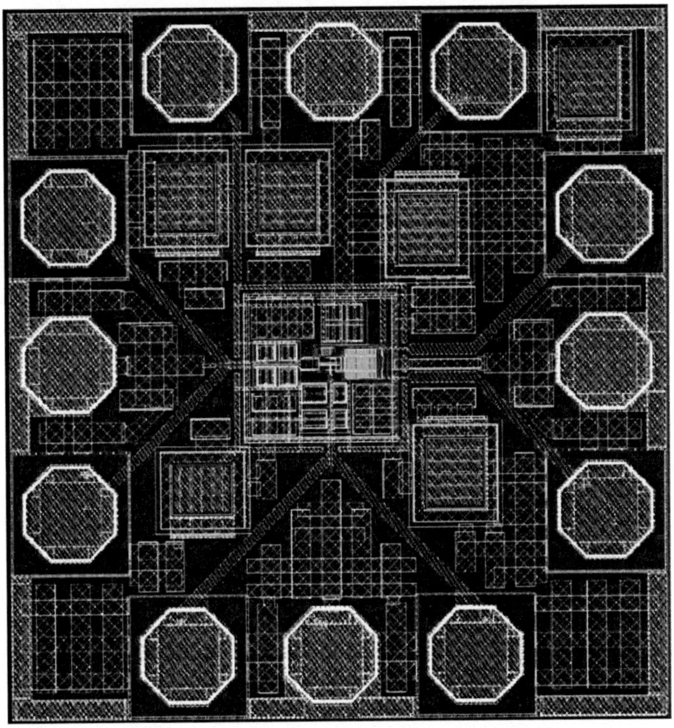

Fig. 7.33 WLAN active mixer complete layout

7.4.2.3 Test Setup and Characterization Results

Fig. 7.34 shows the die microphotography of the active fabricated mixer. The results and some simulations of this mixer and the passive one are exhibited in the next subsection.

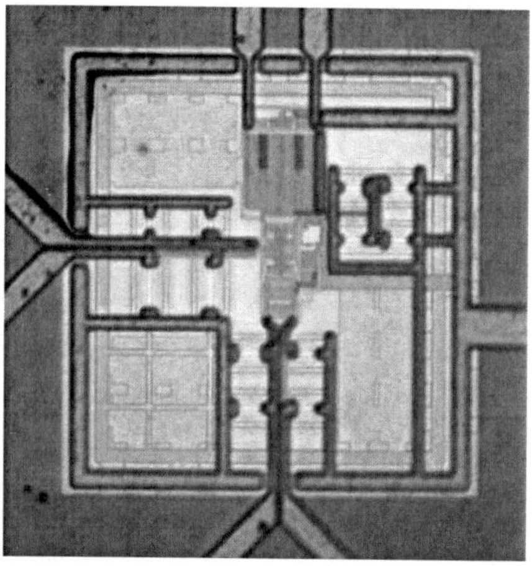

Fig. 7.34 WLAN Active mixer die microphotography

The conversion gain results for the WLAN active mixer are presented in Fig. 7.35. The maximum gain is 9.4dB at 2.5GHz and the mixer 3dB-bandwidth is extended to 7GHz. In the centre of the focused frequency band, at 5.4875 GHz, the measured gain is8dB.The test setup to perform conversion gain measurements is the same used for the DVB-T/H mixer (Fig.7.25).

The noise figure test setup has already been shown in Fig.7.26,however, the high frequency inputs involved in this downconverted noise measurement provoked the unavailability of the HP8971C Noise figure test. Nevertheless, the methodology used by this NF test set can be applied manually. This method is called Y-Factor and the calculation steps have been extracted from the application note 57-2 of Agilent (Agilent 2004). This procedure uses the noise floor obtained by the ESA from the system with and without the Device Under Test (DUT), combining it with and without the noise source. This provides the data used for de-embedding the effects of the system of characterization on the measurements, once the DUT is connected. The only requirement for the usage of this method is a positive gain of the DUT, which is the case of this active mixer.

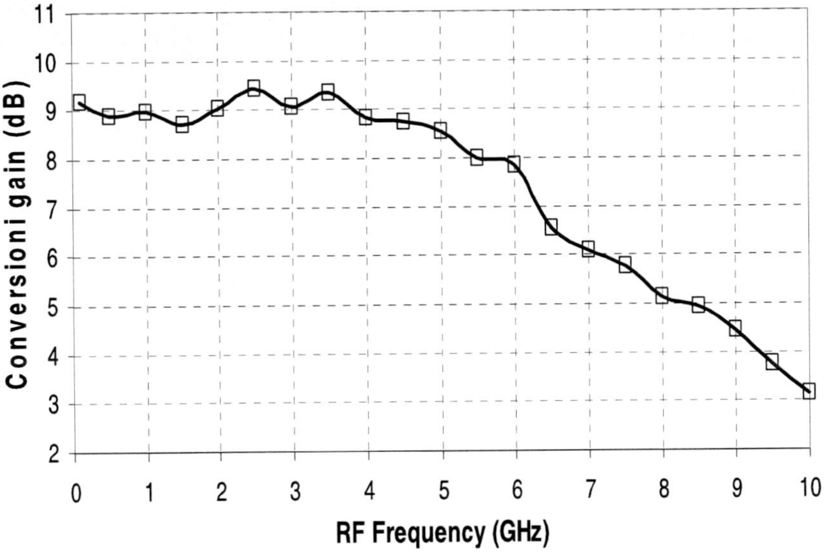

Fig. 7.35 WLAN active mixer conversion gain vs. RF input frequency (GHz)

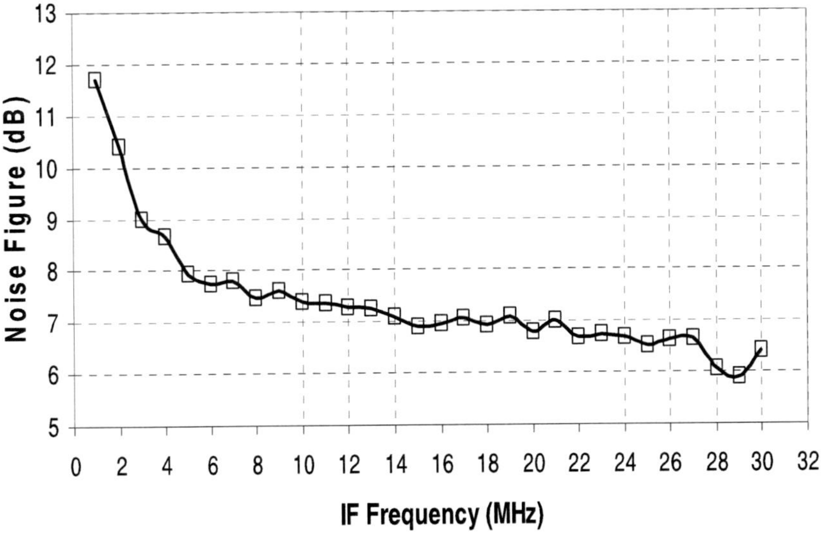

Fig. 7.36 WLAN active mixer NF measured results vs. IF output frequency (MHz)

The Noise figure result is shown in Fig. 7.36. This device exhibited positive conversion gain (Fig. 7.35); therefore, the noise floors detected by this procedure were distinguished. The formulas presented in the application note led to the plot below. Once it overcome the flicker noise corner, the NF laid below 8dB. In the centre of the selected channel, 20MHz, the NF has been 6.8dB (SSB).

Regarding the linearity, Fig. 7.37 outlines the output of the first and the third order harmonic at the output of this device. This measurement has been carried out by a two-tone test, with the test setup shown in Fig. 7.29. Two input tones, at 5.25GHz and 5.245GHz, mixed with an LO tone at 5.27GHz, produces 2 main mixing outputs at 20 and 25MHz. This operation also generates two intermodulation products at 15 and 30MHz respectively. The measured IIP3 is +2dBm.

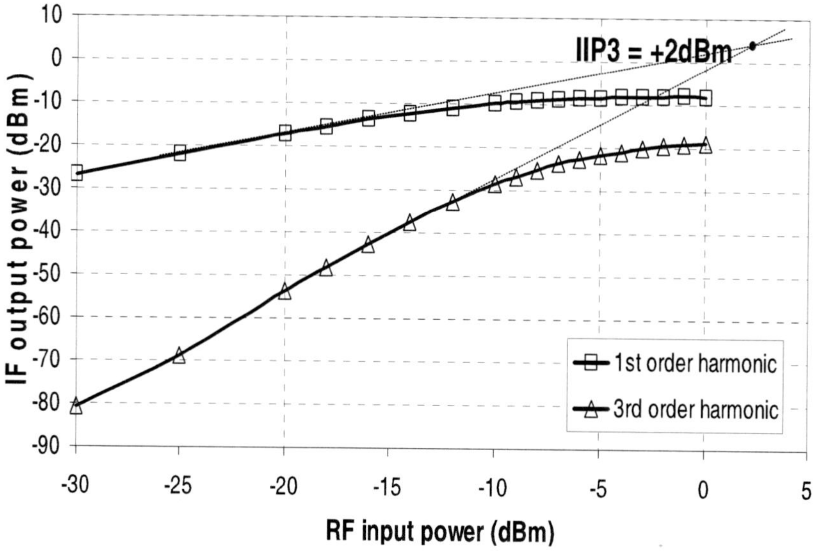

Fig. 7.37 WLAN Active mixer third order intercept point (IIP3)

7.4.2.4 Discussion

The most relevant parameters of the WLAN active mixer's measured performance are highlighted below:

- The conversion gain is 8dB at 5.25GHz. 9.4dB is the maximum value, at 2.5GHz, for a working 3dB-bandwidth of 7GHz
- The IIP3 measured at 5.25GHz is +2dBm
- The SSB Noise Figure is 6.8dB
- The RF rejection at the output port is -27dB and the LO rejection is -55dB

The conversion gain measured at the centre of the 5GHz U-NII 1 and 2 bands meets exactly the specification for this parameter, while the IIP3 is 2dB higher that the required (chapter 3). This is also due to the appropriate polarization of the buffer stage in the weak inversion region.

The NF measured (SSB) is 0.7dB better than the specification. In this case and as it can be observed in Fig. 7.36, the flicker noise corner is situated around 4 MHz; Nevertheless, it is lower than 10 MHz, and hence, suitable for the low IF architecture selected for the front end (section 3.5).

The LO reverse isolation of this active design is -55dB. This provides a suitable isolation, as the higher LO tone fedback to the input of the front end is 23dB lower than the weakest RF signal(IEEE99, IEEE03).

7.4.3 Example 3: Very Low-Power Passive Mixer for Wlan (5GHz U-NII Band)

As mentioned in subsection 7.4.2, the description of the implementation of a low-power passive mixer for WLAN (in the 5GHz U-NII frequency band) is presented in this example. It comprises circuit design issues, layout implementation considerations, the description of the test setup for the measurement of its main parameters, the characterization results as well as a brief discussion about such results and those of its active counterpart (7.4.2) with regard to the stateoftheart.

7.4.3.1 Circuit Design

The selected architecture for the WLAN passive mixer is based on a bridge configuration composed of four switches(subsection 7.2.2.2), as depicted in Fig. 7.38. Considering that the CMOS technology provides excellent switches, the passive mixer foresaw high performances due to the high switching efficiency.

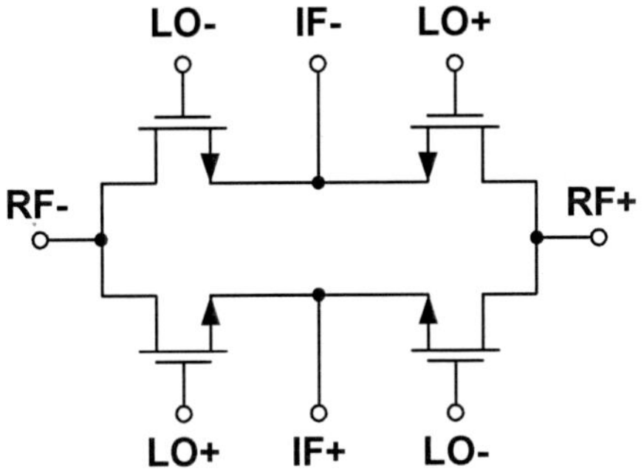

Fig. 7.38 WLAN passive double balanced mixer core configuration

The pairs of identical transistors are distributed diagonally, depending on the LO tone phase. This signal drives the switches and allows the mixing. An appropriate bias point has been selected in order to obtain a suitable combination of noise and linearity performances. It is interesting to note that these two parameters are strongly dependant on the DC bias conditions (Lee 1998). In order to implement the DC operating point keeping the power consumption to a minimum, high value integrated resistors (High Resistive) have been chosen.

A buffer has been implemented in order to allow independent on-wafer block characterization. The conversion loss of the core connected directly to the output load, that typically presents the measurement equipment, were extremely out of range. Thus, a common source amplifier has been placed between the intermediate frequency node and the output of the block. In a fully integrated version of the front end, the size and hence the current consumption of this circuit is reduced.

Since the core current consumption is negligible, the total power consumption of this device is because of the buffer. This added circuitry has to maintain the core NF and IIP3, while preventing the conversion loss, which has been the main challenge for this device. As expressed in subsection 7.4.2.1, the low frequency output of the mixer may benefit from the weak inversion overdrive voltage. The lowering of $V_{gsBUFFER}$ down to 500mV presented a good gain and a lower distortion than in the strong inversion mode, which improved the linearity as mentioned previously. The DC bias point of the buffer is directly dependant on the DC biasing of the RF signal.

7.4.3.2 Layout Implementation

Numerically, the four transistors of the core are the N_L18W500_18_RF RF optimized CMOS transistors of the UMC CMOS 0.18 μm 1P6M technology. The length is the minimum distance of this technology and the Width has been selected for an appropriate Gain / Linearity ratio. In this case, W_{core}=100μm.These four devices are identical, and the common centroid distribution technique has been applied for the layout, as detailed in Fig. 7.39.

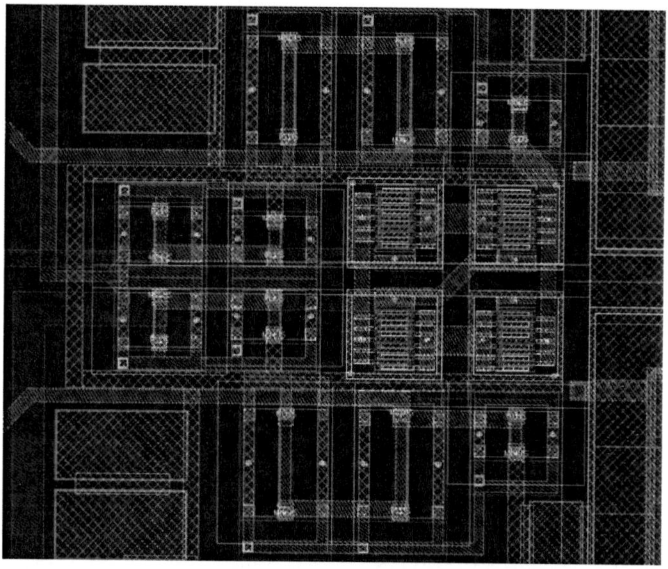

Fig. 7.39 Layout of the WLAN passive mixer core

The transistors of the buffer are also shown in the middle upper part of Fig. 7.39. These devices are N_18_MM devices, with a minimum length (0.18μm) and a width of 1000μm in distributed within 50 fingers, for an appropriate layout shape. The current consumed in each branch is 2.3mA, on the other hand, 288μA wastedthrough the high resistors of the signals' DC biasing. The total current consumed for this circuit is 4.9mA.

The complete layout of the mixer can be observed in Fig. 7.40.

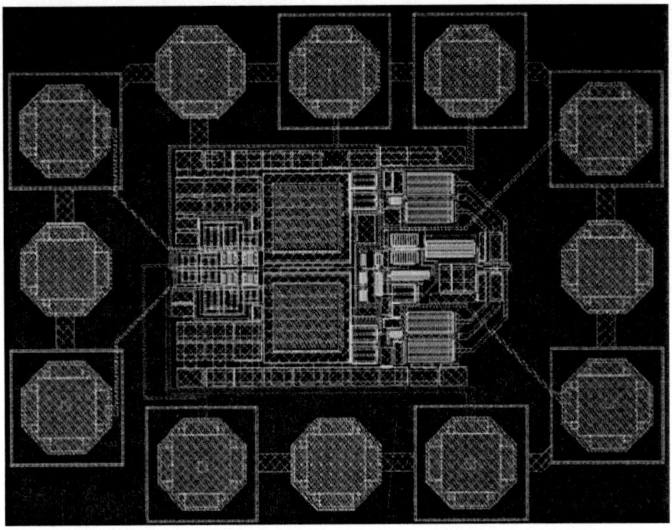

Fig. 7.40 WLAN passive mixer complete layout

7.4.3.3 Test Setup and Characterization Results

The die microphotography of the whole circuit is shown in Fig. 7.41.

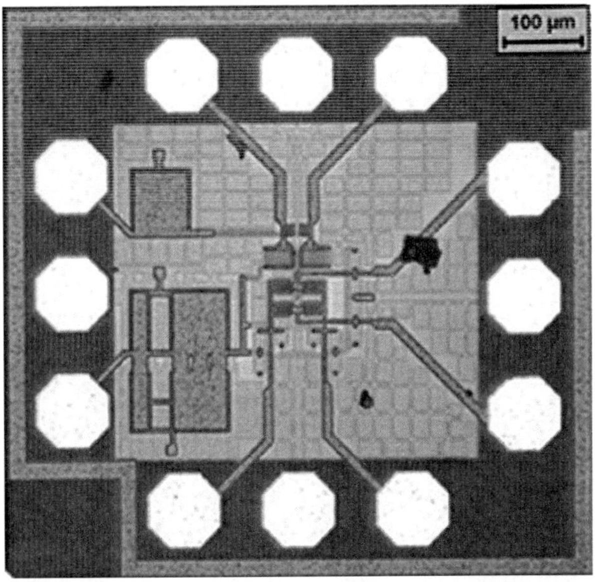

Fig. 7.41 WLAN passive mixer die microphotography

Fig. 7.42 plots the gain of the passive mixer versus the input frequency of the RF signal. This has been performed with the test setup shown in Fig.7.25. The output IF frequency remained fixed at 20MHz. From a minimum loss of 2.1dB, the available cut off bandwidth reached 11GHz. Around the band of interest for the input RF, the gain is centred around -3.5dB.

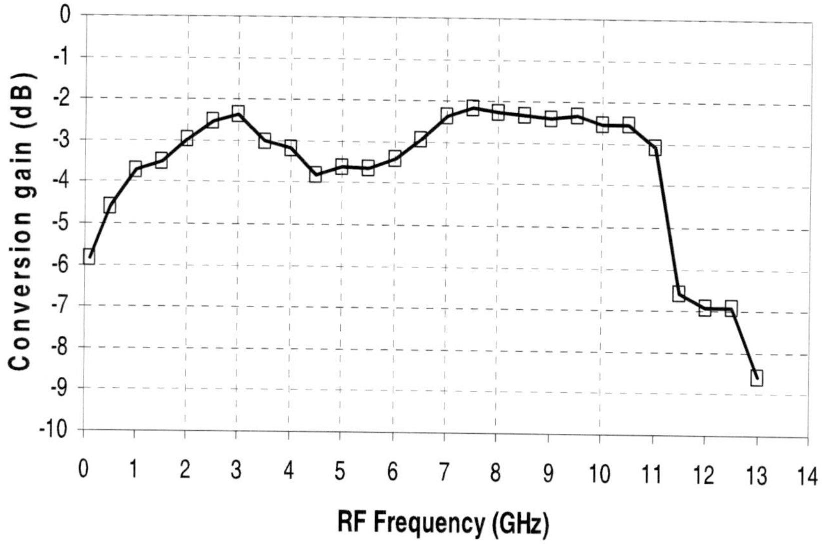

Fig. 7.42 WLAN passive mixer gain measured results versus RF input frequency (GHz)

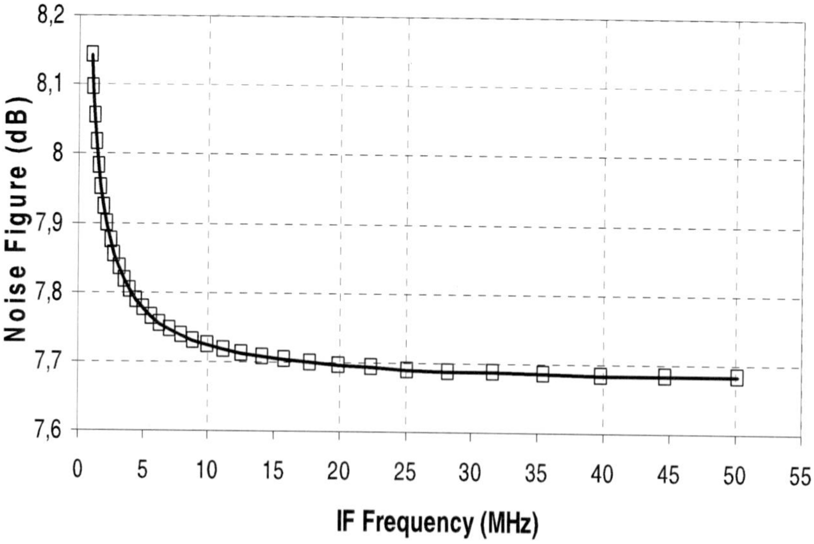

Fig. 7.43 WLAN Passive mixer NF post-layout simulation results

The characterization of the NF for this passive mixer is not possible with the Y-Factor methodology presented in Fig.7.26. As explained in section 7.4.2.3, this device exhibits negative gain (positive conversion loss) and then the noise floor levels distinction is impossible. Fig. 7.43 presents the post-layout simulation using the resistors and capacitors extracted view of the final passive mixer. The NF results laid beyond 8dB once overcame 2MHz. At the centre of the proposed channel (20MHz), the NF has been exactly 7.7dB (SSB).

The resulting IIP3 of this passive mixer, calculated with the measurement setup of Fig. 7.29 is +4.5dBm, as shown inFig. 7.44.

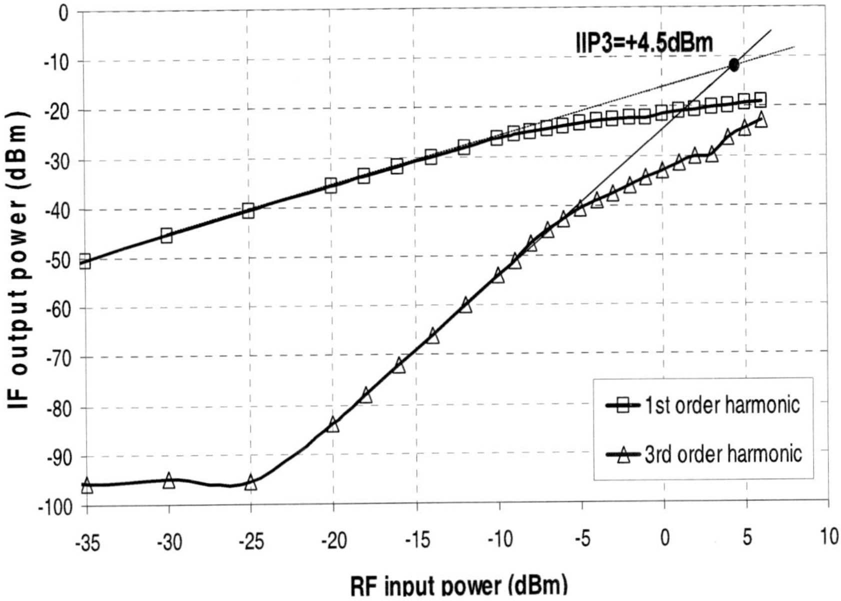

Fig. 7.44 WLAN passive mixer third order intercept point (IIP3)

7.4.3.4 Discussion

The most relevant parameters of the WLAN passive mixer's measured performance are highlighted below:

- The Conversion gain is -3.5dB at 5.25GHz. -2.1dB is the maximum value, at 7.6GHz, for a working 3dB-bandwidth of 11GHz
- The IIP3 measured at 5.25 GHz is +4.5dBm
- The SSB Noise Figure is 7.7dB
- The RF input signal rejection at the output port is -50dB, and the LO rejection is -55dB at the RF port

The measured gain and IIP3 are both 0.5dB higher than the specifications. Moreover, the 3dB-bandwidth of this mixer reaches 11GHz. The high performance of these two parameters is due to the appropriate dimensioning and polarization of the two stages. Furthermore, the sizes selected for these four transistors have minimized the parasitic capacitance, which extends the frequency of operation. On the other hand, the buffer stage works on the weak inversion region and therefore, as expected, the linearity of the circuit increases in comparison with the strong inversion region case.

The measurement of the NF for the passive front end confirms that the NF of the passive mixer accomplishes the required 8dB. The flicker noise corner is situated at 2MHz. It is suitable for the low IF architecture selected for the WLAN front end(section 3.6), centred in 20 MHz.

Regarding the measured isolations, they validate this design for a low IF front end. The LO tone is rejected 55dBat the RF port, which is added to the 39dB reverse isolation of the LNA(subsection6.4.2). In this case, the LO parasitic returned to the input of the LNA is 18dB lower than the weakest RF signal(IEEE99, IEEE03). Therefore, no distortion of the subsequent blocks is possible.

In order to compare the two mixer examples designed for WLAN U-NII (5GHz) presented in subsections 7.4.2 and 7.4.3 with those reported in the bibliography, the Figure of Merit expressed in Eq. (7.28)has been used (Liang 2007).A higher value of this FOM means better general mixer performance. Comparing Eq 7.28 with the Figure of Merit of the LNA(subsection 6.4.2), the only difference is in Eq 7.28 the frequency has not been taken into consideration. Fig. 7.45 presents this comparison. This expression devalues the FOM of the passive mixers with regard to their active counterparts because of the opposite sign of the gain in each case. However, the passive devices present higher IIP3 values and potential lower power consumption.

$$FOM_{MIXER}[dB] = \frac{Gain[abs] * IIP3[mW]}{(NF[abs]-1) * Power[mW]} \qquad \text{Eq. 7.28}$$

The lower current reported for any mixer working in the 5GHz U-NII band is 4.9mA, presented by Sang-sun et al. in (Sang-Sun et al. 2005). This value has been met by the two designs presented in this work. Besides, the FOM of both mixers are higher than in the mentioned paper. The active design of this work has a higher FOM than the passive design, as the gain reaches 8dB. Furthermore, the linearity parameter of this circuit is very close to the IIP3 performed by the passive circuit. However, this FOM has been calculated taking into account the output buffer contribution to the total power consumption. When implementing both mixers in the WLAN front-end (section 3.5) this buffer is not necessary anymore, and therefore mixers' power consumption would be reduced, especially in the case of the active mixer, due to the fact that it's core's current consumption is only 1.9mA.

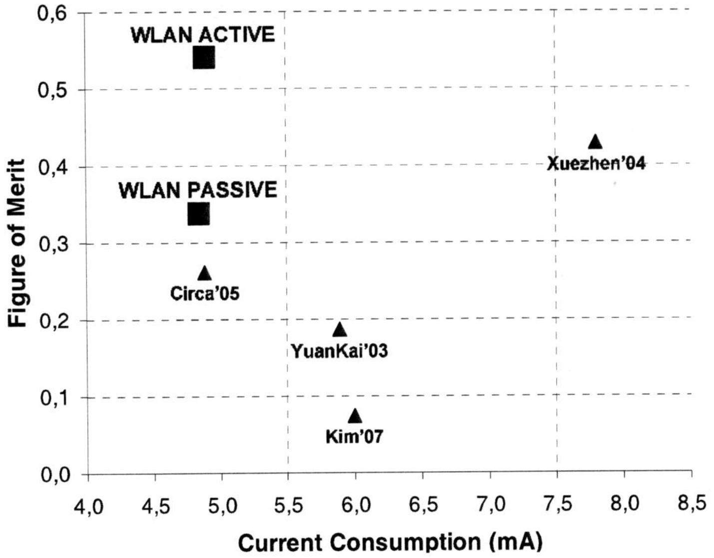

Fig. 7.45 Figure of Merit of CMOS mixers working in the 5 GHz U-NII band

References

(Agilent 2004) Agilent, Application note 57-2: Noise Figure Measurement Accuracy - The Y-Factor Method (2004)

(Alvarado et al. 2008) Alvarado, U., et al.: Low Frequency Noise Optimization in Gilbert-Cell Based Mixers for Direct Conversion (Zero-IF) Receivers. Microwave and Optical Technology Letters 50, 3128 (2008)

(Bautista et al. 2000) Bautista, E.E., et al.: A High IIP2 Downconversion Mixer Using Dynamic Matching. IEEE Journal of Solid State Circuits 35(12), 1934 (2000)

Caverly, R.: CMOS RFIC Design Principles. Artech House, Boston (2007)

(Circa et al. 2005) Circa, R., et al.: Integrated 130nm CMOS Passive Mixer for 5GHz WLAN Applications. In: SBMO/IEEE MTT-S International Conference on Microwave and Optoelectronics, p. 103 (July 2005)

(Clein 2000) Clein, D.: CMOS IC Layout. Concepts, Methodologies and Tools. Newness (2000)

(Coleman 2004) Coleman, C.: An Introduction to Radio Frequency Engineering. Cambridge University Press, Cambridge (2004)

(Crols and Steyaert 1995) Crols, J., Steyaert, M.: A 1.5GHz Highly Linear CMOS Downconversion Mixer. IEEE Journal of Solid State Circuits 30(7), 736 (1995)

(Darabi et al. 2000) Darabi, H., et al.: Noise in RF-CMOS mixers: a simple physical model. IEEE Journal of Solid State Circuits 35, 15 (2000)

(DVB-H_EN302304) DVB-H_EN302304, Transmission System for Handheld Terminals, ETSI standard

(DVB-T_EN300744) DVB-T_EN300744, Framing Structure, Channel Coding and Modulation for Digital Terrestrial Television, ETSI standard

(Egan 2003) Egan, W.F.: Practical RF System Design. John Wiley and Sons, Chichester (2003)
(Fong 2009) Fong, K.: Dual-Band High-Linearity Variable Gain Low-Noise Amplifiers for Wireless Applications. In: ISSCC Symposium, pp. 194–195, 224–225, 463 (February 1999)
(Gilbert 1968) Gilbert, B.: A Precise Four-Quadrant Multiplier with Subnanosecond Response. IEEE Journal of Solid State Circuits 39(4), 365 (1968)
(Gilbert1997) Gilbert, B.: The Micromixer: A Highly Linear Variant of the Gilbert Mixer Using a Bisymmetric Class-AB Input Stage. IEEE Journal of Solid-State Circuits 32, 1412 (1997)
(Gilbert 1998) Gilbert, B.: The Multi-Tanh Principle: A Tutorial Overview. IEEE Journal of Solid State Circuits 33, 2 (1998)
(Go-Young 2000) Jung, G.-Y.: A Low-Noise UWB CMOSMixer Using Current Bleeding and Resonant Inductor Techniques. IEEE Journal of Solid State Circuits 49, 1595 (2007)
(Goldfarb et al. 2003) Goldfarb, M., et al.: Even harmonic double-balanced active mixer for use in direct conversion receivers. IEEE Journal of Solid State Circuits 38(10), 1762 (2000)
(Hastins 2000) Hastins, A.: The Art of Analog Layout. Prentice-Hall, Englewood Cliffs (2000)
(Hsieh and Hung 2007) Hsieh-Hung, Liang-Hung: Design of Ultra-Low-Voltage RF Front-Ends With Complementary Current-Reused Architectures. IEEE Transactions on Microwave Theory and Techniques 55(7), 1445 (2007)
(IEEE99) IEEE, WLAN 802.11a: Wireless LAN Medium Access Control (MAC) and Physical Layer (PHY) Specifications. High-speed Physical Layer in the 5 GHz Band (1999)
(IEEE03) IEEE, Part 11: Wireless LAN Medium Access Control (MAC) and Physical Layer (PHY) specifications Amendment 4: Further Higher Data Rate Extension in the 2.4 GHz Band, IEEE, SA Standard Board (2003)
(Karanicolas 1996) Karanicolas, A.N.: A 2.7 V 900 MHz CMOS LNA and mixer. IEEE Journal of Solid State Circuits 31, 50 (1996)
(Kim et al. 2007) Kim, J.-H., et al.: Design of Reconfigurable RF Front-End for Multi-Standard Receiver Using Switchable Passive Networks. Analog Integrated Circuits and Signal Processing 50(2), 81 (2007)
(Lee 1998) Lee, T.: The Design of CMOS Radio Frequency Integrated Circuits. Cambridge University Press, Cambridge (1998)
(Lee et al. 2000) Lee, S.G., et al.: Current-reuse bleeding mixer. Electronics Letters 36, 696 (2000)
(Liang 2007) Liang: The Exploration of Spectrum Monitor Architecture for Cognitive Radio, University Of Southampton, PhD Dissertation (2007)
(Liu 1998) Liu, E.A.W.: BSIM3v3.2 MOSFET Model Users Manual, The Regents of the University of California (1998)
(Maas 1987) Maas, S.A.: A GaAsMESFETmixer with very low intermodulation. IEEE Transactions on Microwave Theory and Techniques MTT-35, 425 (1987)
(Ming-Feng et al. 2004) Huang, M.F., et al.: A CMOS Even Harmonic Mixer with Current Reuse for Low Power Applications. In: Proceedings of the 2004 International Symposium on Low Power Electronics and Design (August 2004)
(Phan et al. 2005) Phan, T.A., et al.: A High Performance CMOS Direct Down Conversion Mixer for UWB System. IEICE Transactions on Electronics 88(12), 2316 (2005)

(Razavi 1998) Razavi, B.: RF Microelectronics. Prentice Hall PTR, Englewood Cliffs (1998)

(Razavi 2001) Razavi, B.: Design of Analog CMOS Integrated Circuits. McGraw-Hill, New York (2001)

(Sang-Sun and Hyung-Joun 2005) Yoo, S.-S., Yoo, H.-J.: A CMOS Multi-Standard Mixer for WCDMA, Wi-Bro and 802.11a/b/g. In: Microwave Conference Procedings, Asia-Pacific Conference Proceedings, APMC (2005)

(Shaikh and DeGroat 2006) Shaikh, K.A., DeGroat, J.: A 5GHz CMOS Low Power Down-Conversion Mixer for Wireless LAN Applications. In: Proceedings of the 5th WSEAS International Conference on Circuits, Systems, Electronics, Control and Signal Processing, p. 26 (2006)

(Sedra and Smith 2004) Sedra, A., Smith, K.C.: Microelectronic Circuits. Oxford University Press, Oxford (2004)

(Terrovitis et al. 1999) Terrovitis, M.T., et al.: Noise in Current-Commutating CMOS Mixers. IEEE Journal of Solid State Circuits 34, 772 (1999)

(Xuezhen et al. 2003) Wang, X., et al.: A novel 1.5 V CMFB CMOS down-conversion mixer design for IEEE 802.11 A WLAN systems. In: Proceedings of the 2004 International Symposium on Circuits and Systems, ISCAS, vol. 4, p. 373 (2004)

(Xuezhen and Webster 2004) Xuezhen, W., Weber, R.: A Novel Low Power Low Voltage LNA and Mixer for WLAN IEEE 802.11a Standard. In: Silicon Monolithic Integrated Circuits in RF Systems, p. 231 (2004)

(YuanKay et al. 2003) Chu, Y.-K., et al.: 5.7GHz 0.18um CMOS Gain-Controlled LNA and Mixer for 802.11a Applications. In: Radio Frequency Integrated Circuits (RFIC) Symposium, p. 221 (2003)

(Zencir et al 2002) Zencir, E., et al.: A Low-Power CMOS Mixer for Low-IF Receivers. In: Radio and Wireless Conference, RAWCON, p. 157 (2002)

8 Phase Locked Loop (PLL) Design

The last building block covered in the book is the Phase Locked Loop (PLL), virtually used in every integrated communication front-end. The complexity of this system is high; therefore the first section is dedicated to the basic concepts related to both PLL system architectures and basic components. The architecture of the classical phase locked loops used in RF IC designs are presented in that first section. Nevertheless, from the power consumption point of view, the decision on the architecture of the whole PLL is an important point, but the internal design of each block is also a key issue.

The three main and most challenging blocks are explained with a greater level of detail in subsequent sections: Phase Frequency Detector (PFD) is first described in section 8.2, then section 8.3 presents the design constraints related to the Voltage-Controlled Oscillator (VCO), and High Frequency Divider (HFD) is deeply analyzed in section 8.4. The two last blocks (VCO and HFD) are crucial in the whole power consumption of this complex circuit as these have to work in the high frequency bands of the application. Design examples are shown for these two blocks in section 8.5.

8.1 Frequency Synthesis Fundamentals

This section presents a basic review of PLL fundamentals. The introduction is dedicated to presenting a building block diagram and the basic formulas of the most common PLL architectures, and the subsequent subsection goes into detail in the two most important architectures for RF integrated circuits: integer-N and fractional.

8.1.1 Introduction to PLL

The most suitable integrated frequency synthesizer for RF applications at working frequencies of various GHz is the Phase Locked Loop (PLL). In Fig. 8.1, the block diagram of a PLL, also called an indirect frequency synthesizer, can be seen. The working principle of this synthesizer consists of continuously correcting the phase difference that exists between the two periodic input signals to the loop, i.e. F_{ref} and F_{div}. When there is no difference, it is said that the loop is locked, and the output frequency of the synthesizer (F_{out}) is a multiplication of the reference frequency F_{ref} by a variable number (N).

U. Alvarado et al.: Low Power RF Circuit Design in CMOS Tech., LNEE 104, pp. 179–236.
springerlink.com

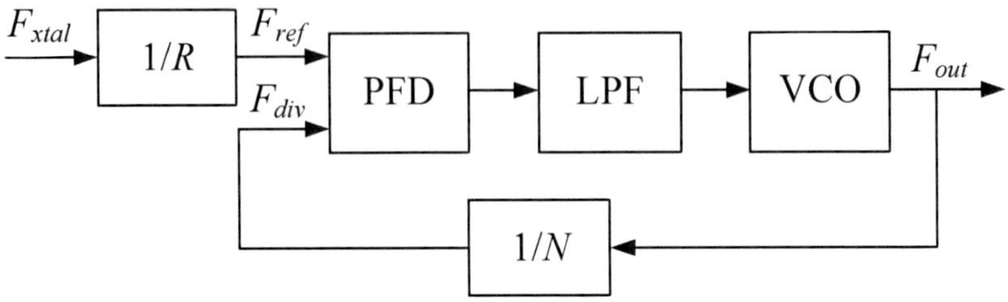

Fig. 8.1 Phase-locked loop

The output signal of the VCO is frequency-divided and the resulting signal (F_{div}) is compared with the reference frequency (F_{ref}) in the phase detector (PD), generating a proportional signal to the phase difference that exists between the two input signals. Generally, in CMOS PLLs together with the PD a charge pump (CP) is usually integrated, being responsible for transforming the PD output signal into a source of current pulse train, dependent on the output. Further discussion on these blocks can be found in section 8.2.

The CP output signal is filtered using a low-pass filter (LPF), which converts the current pulse train into a continuous voltage to control the VCO. Indeed, it acts as a transimpedance network to perform the current to voltage conversion as well as the filtering. The frequency response of this LPF impacts the overall PLL performance and its stability. Furthermore, this filter attenuates the undesired spurious emissions and determines the bandwidth of the loop, a parameter that influences the total noise of the PLL.

As a rule of thumb, even in an Integrated Circuits design context, the elements that compose the LPF can overpass the technological limits given by the foundries and are made of discrete elements. This is a common issue for the researchers in charge of the modern PLLs design.

When the loop is locked, the two inputs to the phase detector maintain a relation of constant phases and as a consequence the same frequency. The synthesized frequency can be obtained using the following formula:

$$F_{out} = \frac{N}{R} \cdot F_{xtal} \qquad \text{Eq. 8.1}$$

In this way, possible variations in the input phase signal from the VCO will be transmitted at the input and corrected by means of a phase detector and low-pass filter.

8.1.2 PLL Architectures

Depending on the division value N, this type of synthesizer is usually classified as an integer synthesizer or a fractional synthesizer. The following sub-subsections briefly explain the operation of each one of these. The derivation of the transfer function has been omitted in order to simplify the text, but they can be found in some of the references listed at the end of the chapter (Craninckx and Steyaert 1998) (Rohde and Newkirk 2000).

8.1.2.1 Integer-N PLL Architecture

The architecture of this synthesizer is presented in Fig. 8.2, where the input frequency divider R has been omitted for simplicity. This type of synthesizer generates a frequency F_{out} that can be calculated according to Eq. 8.2, where N varies in unit stages from N_L to N_H.

$$F_{out} = N \cdot F_{ref} \qquad \text{Eq. 8.2}$$

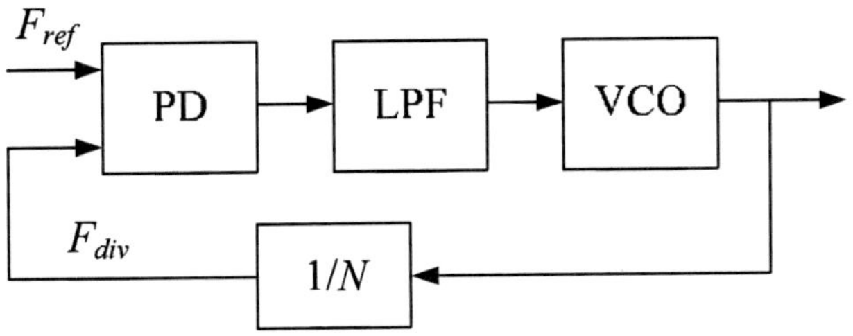

Fig. 8.2 Integer-N PLL architecture

The key point in this architecture is that the reference frequency (F_{ref}) must coincide with the spacing between channels in order to be able to make a selection from these. This is due to the fact that N has to be an integer in this architecture and consequently the simplest divider is then needed. Furthermore, no harmonics caused by lower frequency tones disturb the comparison process and the subsequent blocks. In this way, when N takes the value N_L the inferior channel is selected. Later, by means of the variation of N the rest of the channels are selected successively until the superior channel is reached which is tuned when N reaches the value N_H. Therefore, the frequency divider in Fig. 8.2 must provide a variable division ratio given in Eq. 8.3.

For example, for a 10MHz channel spacing, a crystal quartz of 10MHz would make *k* be from 0 to M, instead of from 0 to 2M as in the case of a crystal quartz of 5MHz. Moreover, for this last hypothesis, the odd harmonics of 5MHz are also present in the signal path that can result in errors of the control of the VCO.

$$N = N_L + k, k = 0,1,\ldots,M \qquad \text{Eq. 8.3}$$

An example of this type of divider is the Pulse-swallow frequency divider, illustrated in Fig. 8.3. This divider consists of a prescaler, a program counter and a swallow counter. Before briefly describing the way it works three observations need to be made:

- the prescaler divides the input by either M or M+1, depending on the logic state of the control line of the module

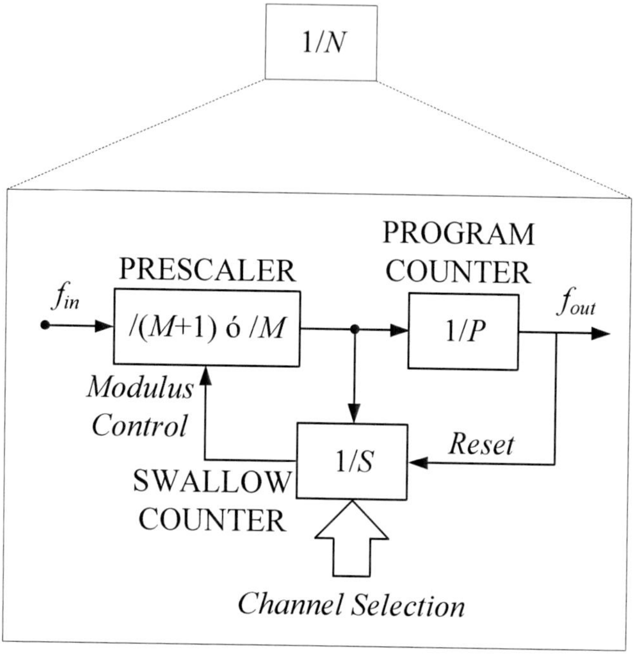

Fig. 8.3 Pulse-swallow frequency divider

- the program counter always divides the input of the prescaler by P
- the swallow counter divides the gate of the prescaler by S, where S is determined by the digital channel selection input and can vary from 1 to the maximum number of channels. The swallow counter also has a reset input.

The output frequency of this divider can be calculated using Eq. 8.4.

$$f_{out} = \frac{f_{in}}{MP + S} \qquad \text{Eq. 8.4}$$

When the circuit is in the reset state, the prescaler divides by M+1. The output of this prescaler is divided as much by the program counter as by the swallow counter until this last one is complete, that is to say until S pulses have been counted. At this point, after (M+1)·S cycles in the main input, the swallow counter changes the state of the module line control, making the prescaler divide f_{in} by M. Before this change is evident the program counter has already counted a total of S pulses. Following this change, the prescaler and the program counter keep on dividing until this last one is complete. Given that the program counter has already counted S pulses, P-S cycles are required at its input and therefore (P-S)·M pulses at the main output in order to reach its end of count. Therefore, the main output generates a complete cycle every (M+1)·S+(P-S)·M cycles at the input, that regrouping terms results in one every P·M+S cycles. The operation is repeated after the swallow counter is restarted.

The main characteristic of this architecture is its simplicity, which has made it a common choice in the implementation of integrated frequency synthesizers for many decades (Razavi and Lam 2000), (Valla et al. 2005), (Herzel et al. 2003), (Rategh et al. 2000), (Zhang 2003), (Behzad et al. 2003), (Vassiliou et al. 2003), (Leung and Luong 2004), (Zargari et al. 2002), (Ahola et al. 2004).

However, this architecture also has some disadvantages. On the one hand, it is important to point out the appearance of spurious emissions in the output signal at a distance from the carrier equal to the reference frequency (F_{ref}), as illustrated in Fig. 8.4.

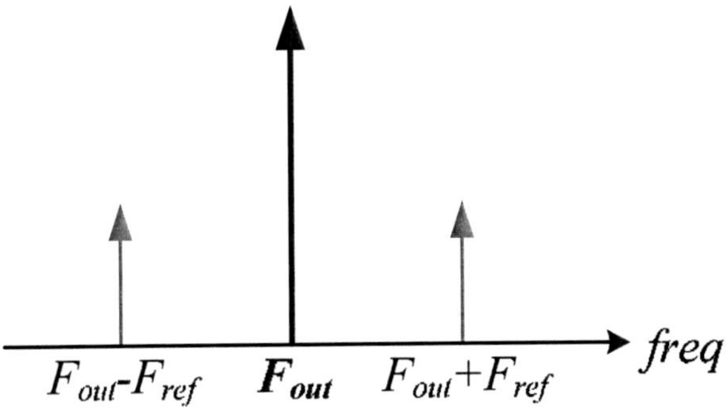

Fig. 8.4 Spurious emissions due to the reference frequency

These spurious emissions can be attenuated with an appropriate loop filter design. On the other hand, the reference frequency needs to coincide with the spacing between channels in order to be able to carry out the tuning of these. This requires a limitation in the loop bandwidth given that its determination is a trade-off between the time it takes to establish the channel (lock time) and, spurious emissions and phase noise at the output of the synthesizer. As the spurious emissions appear at a distance F_{ref} from the carrier, a bandwidth far superior to this frequency make these emissions attenuate little but also make the system quicker and vice versa. The same is true for the phase noise. Therefore, a trade-off between phase noise at the output and the speed of tuning is necessary for the determination of the bandwidth.

8.1.2.2 Fractional PLL Architecture

As has been outlined before, the main disadvantage of integer-N architecture is that the reference frequency has to coincide with the separation between channels, which limits the bandwidth of the loop. On occasions the separation of channels is too small for integer-N architectures to be used, making it necessary for the reference frequency to be higher than the said separation. On these occasions, channel selection is carried out by varying the output frequency by a fraction from the reference frequency. This is the basis of fractional architecture synthesizers.

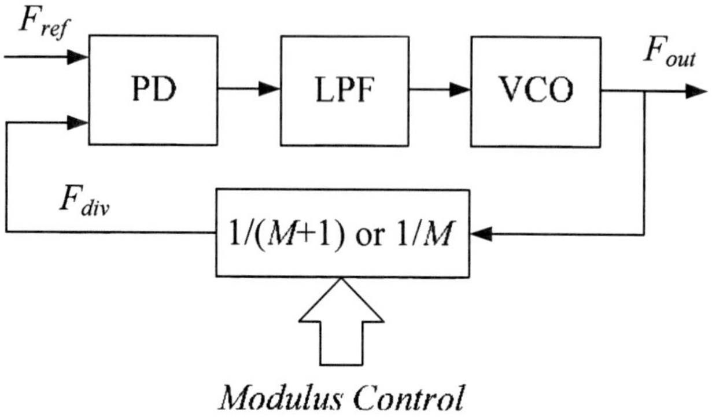

Fig. 8.5 Fractional architecture using a dual-modulus divider

The fractional architecture block diagram is shown in Fig. 8.5 (Razavi 1998). This architecture substitutes the frequency divider from integer-N architecture for a dual-modulus prescaler. If AVCO output pulses are divided by this prescaler by the value M and B pulses by the value M+1, then the equivalent ratio of division (N_{eq}) can be obtained from Eq.8.5. This value can vary between M and M+1 in fine steps using the correct choice of values A and B.

$$N_{eqt} = N.f = \frac{A+B}{\frac{A}{M}+\frac{B}{M+1}} \qquad \text{Eq. 8.5}$$

The resulting division ratio is sometimes denoted as N.f, where the point refers to a decimal point and N and f represent the integer and fractional parts of the division ratio respectively.

As previously mentioned, this type of architecture mitigates the need for the reference frequency to coincide with separation between channels. In this way, with F_{ref} in the range of tens of MHz, the loop bandwidth of a fractional synthesizer can be in the order of just a few MHz, something that produces a quick transient response together with a suppression of the VCO phase noise within this bandwidth. Furthermore, the reduction of the division ratios (when F_{ref} is increased) reduces the effect of the PD phase noise.

But this architecture presents the critical disadvantage of fractional spurious emissions. In the same way that integer-N, fractional architecture presents spurs at fractional frequencies to those of the reference. For example, if the output frequency of the VCO is equal to $(N+\alpha)\cdot F_{ref}$, where N represents the integer part of the divisor and α the fractional part, the fractional spurs are introduced at frequencies$\alpha\cdot F_{ref}$, $2\alpha\cdot F_{ref}$, etc. The problem with the fractional spurs is quite serious given that their levels are superior to those of the spurs that appear in integer architecture (typically between 20 and 30 dB below that of the carrier), the use of additional circuitry to compensate is usually necessary. These additional circuits can introduce considerable phase noise in the synthesizer output signal, and of course, add more circuit complexity.

8.2 Phase-Frequency Detector (PFD) Design Constraints

A phase frequency detector / phase detector generates an output signal proportional to the phase and / or difference that exists between its two input signals. One of the input signals is fixed, with very stable frequency and generally generated by a quartz crystal. The other input signal can be variable, less stable and comes from the output of the oscillator after passing through the frequency divider. The function of the phase frequency detector / phase detector within the loop is to correct the excess of phase that exists between two inputs and to lock the frequency by means of a slight variation in the VCO voltage control.

Currently four main types of phase frequency detector / phase detectors are used:

- Analogical or multiplying phase detectors that are based on the multiplication of two sinusoidal signals from the same frequency.
- The first sequential circuits which operate with the information provided by the zero crossings of the input signals is the logic gate OR-exclusive.
- The flip-flop based phase detector, which is also a sequential circuit that uses the zero crossings.
- The fourth category is a sequential circuit that also provides a signal dependent to the frequency: this is a PFD, in comparison to the three first types that are only PD. These blocks are aimed to lock the loop when the PLL output signal is unlocked.

8.2.1 Multipliers

If the two phase detector inputs are sinusoidal, a mixer or multiplier can be used as a phase detector. In order to clarify this point, let us consider two signals as shown in Eq. 8.6 and Eq. 8.7. If we use these signals as inputs of a balanced mixer phase detector, the output-generated signal would follow Eq. 8.8.

$$v_1 = A_1 \sin(\omega_1 t + \theta_1) \quad \text{Eq. 8.6}$$

$$v_2 = A_2 \sin(\omega_2 t + \theta_2) \quad \text{Eq. 8.7}$$

$$v_d = A_d \left\{ \sin\left[(\omega_1 - \omega_2)t + \theta_1 - \theta_2\right] + \sin\left[(\omega_1 + \omega_2)t + \theta_1 + \theta_2\right] \right\} \quad \text{Eq. 8.8}$$

Where the amplitude of the signal resulting from the product (A_d) is given by

$$A_d = \frac{A_1 A_2}{2} \qquad \text{Eq. 8.9}$$

In phase-locked loop condition, the two angular frequencies are the same and the DC component at the phase detector output equals $A_d \sin(\theta_1 - \theta_2)$, which is proportional to the phase difference for small values of $\theta_1 - \theta_2$.

This phase detector is especially useful in applications where the reference frequency is too high for other solutions. The main disadvantage of this type of phase detector is the high number of undesired frequency components generated at the output, which have to be attenuated by the loop filter. The most disturbing is the component situated at the frequency sum $\omega_1 + \omega_2$, which in locked condition will be at twice the reference frequency, and consequently is directly added to the first harmonic spurious peak of the reference tone.

That last disadvantage and the fact that the mixer or multiplier design is complex, advise against the use of this type of PFD from a low power point of view.

8.2.2 Exclusive-OR Logic Gate and Flip-Flops

An Exclusive-OR (X-OR) logic gate or a flip-flop can be used as a phase detector, as demonstrated in Fig. 8.6 and Fig. 8.7. These figures present the functioning of these logic structures when two slightly out of phase signals AandB arrive at each one of their inputs.

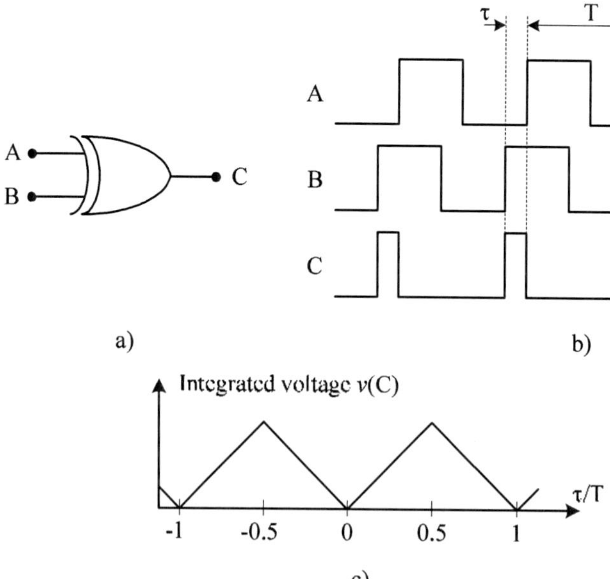

Fig. 8.6 X-OR phase detector

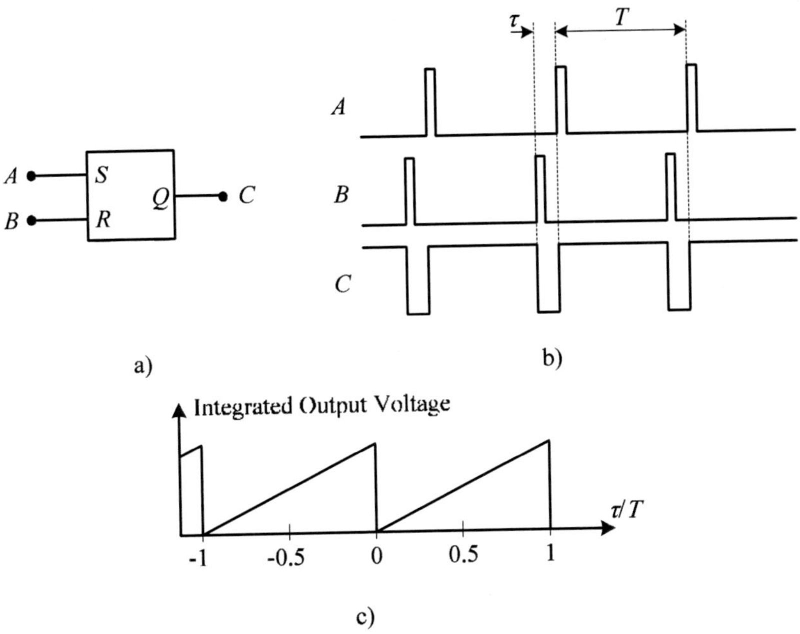

Fig. 8.7 Flip Flop phase detector

Firstly, for the X-OR logic gate, the two input signals A and B are shown in Fig. 8.6b, together with the output signal C. Its relative mean value is proportional to the phase difference on a half cycle range, as shown in Fig. 8.6b. The operating point of this phase detector must be chosen in the middle of the linear range of its transfer function shown in Fig. 8.6c, which corresponds to a phase difference of 90° between its inputs. From this operating point the phase difference between A and B is allowed to shift 90° in both directions, achieving a maximum working range of 180°, with no impact on the duty cycle of A/B.

Secondly, in the set-reset (S-R) flip-flop illustrated inFig. 8.7, a narrow pulse train in both inputs A and B activate and deactivate output C. The average value of C presents a saw-tooth waveform, with a linear range of a complete cycle (360°).

The power consumption of these logic gates comes from two sources: the dynamic power consumption is due to the switching behavior, whereas the static power consumption is caused by the leakage currents. The observation on this power consumption from the functional point of view sets that it depends exclusively on the operating frequency, and consequently it would be greater for higher input frequencies. It finally results in a low-power consumption as it is mainly composed of the dynamic addend. Nevertheless these structures are only available for digital phase detectors, and then, need an evolution to be included in analog integrated circuits designs.

8.2.3 PFD/CP

The PFD/CP (Phase-Frequency Detector with Charge Pump) is a sequential phase detector with the ability of also locking the frequency. It is mostly used for the implementation of locked loops, due to 3 beneficial characteristics:first, it presents an input linear range of ±360°; secondly, it acts as a phase detector while the loop remains locked and provides a signal proportional to the difference in frequency that contributes to lock PLL. And finally, the circuit schematic is compatible with CMOS integrated circuit technologies. The block diagram of this type of detector is shown in Fig. 8.8.

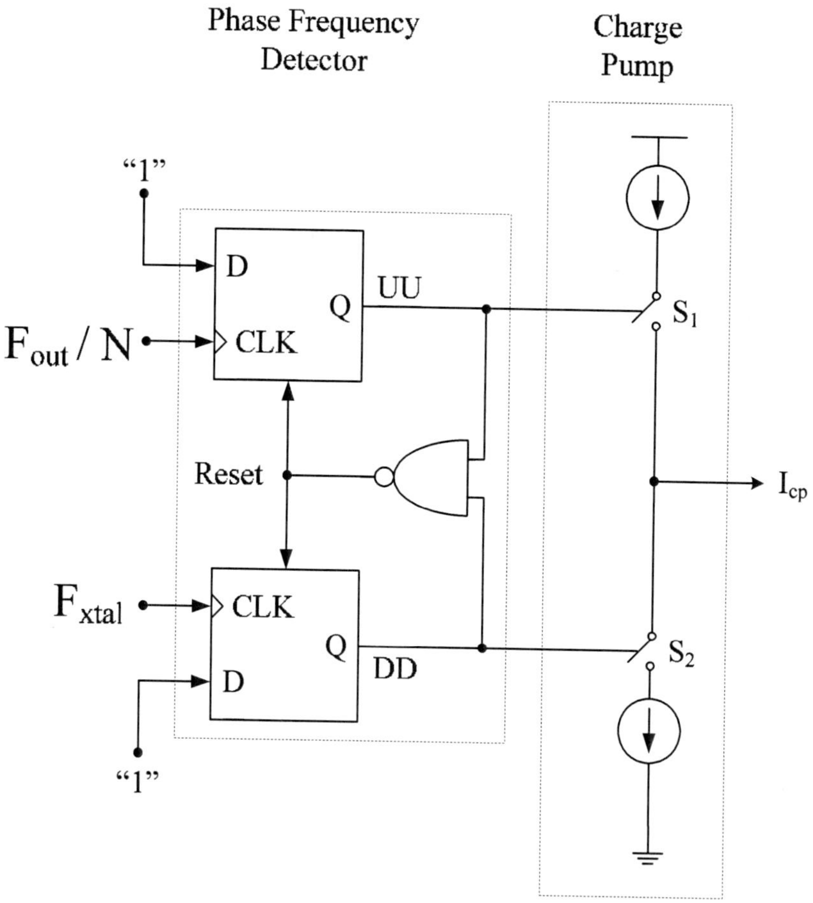

Fig. 8.8 Phase-Frequency detector with charge pump

As his name indicates, the circuit consists of two distinct parts:

- The Phase-Frequency Detector (PFD), responsible for generating voltage pulses of a width proportional to the phase and frequency difference between its inputs.
- The Charge Pump (CP), responsible for delivering a charge proportional to this phase/frequency difference.

A simple implementation of the PFD, using type D flip-flops, is shown inFig. 8.8. Before we discuss the complete system, it is useful to remember the operation of D flip-flops: when the clock input is at zero it stays invariable, and when the clock input changes at a high level the data input is copied to the output. Independently of the clock and the data input, when the reset input is activated the output of the flip-flop goes to a low level.

The PFD consists of two D flip-flops and a NAND gate. As an initial condition, it is assumed that both outputs (UU and DD) are reset (at a low level). When the first edge arrives by either of the two clock inputs, the flip-flop in question will copy the input data (always connected to a high level) into its output D. On the other hand, the other flip-flop will continue as at the beginning (with the output at a low level) until the first positive clock edge arrives. When this happens it will copy the input data D (also at a high level) to its output. At this point the two outputs of the flip-flops happen to be at a high level, something that will cause the logic gate AND (or NAND) to activate its output and reset both flip-flops. From this point everything starts again from the beginning.

It is important to point out that during the time that the NAND gate takes to reset both flip-flops a small pulse at the output of one of the flip-flops will appear (in the flip-flop in which the rising edge arrives later), which has not been represented due to its short duration (of the order of few nanoseconds).

The working principle of the CP can be described as follows: if the line UU is active, the charge pump generates positive current pulses (acting as source); on the contrary, if DD is active these current pulses will be negative (drain). In normal operation, one of the inputs of the PFD is connected to the reference crystal (F_{xtal}), while the other comes from the oscillator output after being divided by the divider (F_{out}/N). The output current drives the low-pass filter of the loop, which is responsible for transforming the pulse train into a voltage value.

This type of phase detector, despite being the most used in the implementation of locked loops (Leung and Luong 2004), (Herzel et al. 2003), (Rategh et al. 2000), (Zhang et al. 2003), (Vassiliou et al. 2003), presents non-ideal effects that may degrade the PLL overall performance, as a consequence of a modulation in the VCO line of control. The most important of these problems is dead zone, which has also implication on the power consumption of the PFD/CP.

The dead zone of a PFD/CP phase detector is the phase difference between its two inputs for which the charge pump does not inject current in the loop filter. Further explanations can be found in the references, but, basically, concerning the current consumption, this dead zone provokes the rising of the jitter at the output of the loop, that is, an increase in phase noise of the PLL.

One of the most used methods to eliminate the dead zone consists of including an even number of inverters in the reset of the PFD flip-flops that do not change its logic state, but give rise to a delay in the signal large enough to cancel it. In this way, an ideal locked loop situation with null phase difference, the two outputs of the PFD generate narrow voltage pulses of the same width that activate the charge pump and inject current pulses of the same amplitude in the loop filter. As a consequence of this, the dead zone is eliminated.

But while the dead zone is eliminated, the reset pulses also introduce negative effects. The most important are due to the delay that exists between the pulses from UU and DD (see Fig. 8.8), the difference in absolute value between the charge and discharge currents from the charge pump and the finite capacitance seen from the current source drain of the charge pump.

As the charge pump needs both to supply and drain current pulses, one of the two switches of the charge pump must be implemented with PMOS transistors. Due to this, it is necessary to introduce an inverter at the corresponding PFD output to ensure the correct commutation. This gate creates a small delay in one of the two commutation paths, which gives rise to two small current pulses of the same amplitude and width, but with different sign and phase between each other. This does not modify the VCO voltage control because the total charge injected is zero, but it provokes a periodic jitter in this voltage, creating undesired reference spurs.

The second negative effect comes from the difference in absolute value between the charge and discharge currents from the charge pump. If this occurs, in each moment of the comparison the pulses from the PFD output in locked condition give rise to a positive or negative charge injection that displace the VCO voltage control by a determined value. In order to solve this, the PLL introduces a phase error between the input and the output in a way that the total current injected by the charge pump in each cycle is zero. In this way, one of the current pulses will have bigger amplitude but will be narrower. It is useful to mention three important issues related to this effect: The VCO voltage control will also experience a periodic jitter. The mismatching between the charge and discharge currents depends on the voltage of the output from the charge pump. The mismatching of current injection between the two transistors that act like switches in the charge pump and the feed through effect from the clock also increase the phase error and the jitter in the VCO tuning voltage.

The second effect is also another fundamental cause of reference spurs. The third and last effect is known by the name of charge sharing. The parasitic capacities seen from the current source drains contribute notably to the generation of this effect. Deeper explanation and graphical representations of these effects can be found in (Quemada et al. 2009).

As last realistic behavior, mismatch between the current sink and current source is a source for output ripple. In the ideal state, when PLL is locked, the net current needs to be zero. But, because of the current mismatch, the PLL must intentionally create a phase difference to reduce the net current, and this can appear as spurious in the frequency output. A solution is to use an NMOS-only design for the current source/sink. The added benefit of this approach is faster switching speed by steering the bias current, at the cost of added current consumption since the bias current is not shut off when not used.

8.3 Voltage-Controlled Oscillator Design Constraints

Low power design often requires direct conversion architectures, such as low-IF or zero-IF. Any of these two possibilities needs a low power, low phase noise voltage control oscillator (VCO) in the frequency synthesizer. This section is focused on low power considerations applied to the practical modern conception of this device. Fulfilling the specifications required for each application (output power, phase noise, frequency range) the design of this block should be completed with a deeper step into the power consumption consideration. A conscious design leads moreover to an improvement in the results obtained by the classical decisions and assumptions. The increase of the quality factor of the passive elements is one of the key points, followed by an accurate design of the architecture scheme. Furthermore, lower current consumption provides higher oscillation frequencies and facilitates higher frequency ranges, which follow the trends of modern wireless and wideband communication standards. In section 8.5 a couple of VCO examples are described with the above considerations.

8.3.1 Functional Description

First, the basics on the functional description of the Voltage Controlled Oscillators are here presented. This type of oscillator is based on a LC-tank circuit that generates a periodic output signal when it resonates. The oscillator, understood as a single output circuit (see Fig. 8.9), is divided into the following three elements:

- Active circuit
- Resonating circuit
- Positive Feedback path

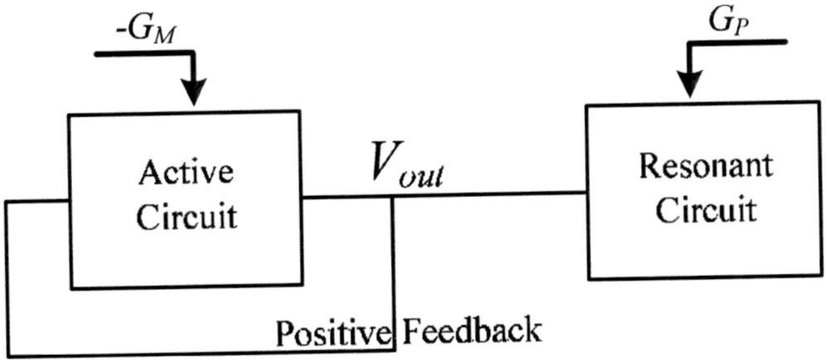

Fig. 8.9 Tank simplified diagram

The LC-tank resonant circuit determines the oscillation frequency at the output according to Eq. 8.10 and the active circuit is an amplifier that, by means of regenerative feedback, adds to the tank the energy needed to compensate its losses, and allow oscillation to be maintained.

If both the inductor and the capacitor were ideal, that is to say without losses or any parasitic elements, a small signal would be sufficient for the oscillator to start to oscillate at the frequency (f_{osc}) given by Eq. 8.10:

$$f_{osc} = \frac{1}{2 \cdot \pi \cdot \sqrt{L \cdot C}} \qquad \text{Eq. 8.10}$$

However, as both the inductor and capacitor exhibit parasitic effects, the LC-tank presents signal loss or attenuation, as well as a shift in desired oscillation frequency (usually downward) caused by these parasitics. A conductance in parallel to the tank (G_p) can be used to represent the losses from the inductor and the capacitor, and the output resistance from the active circuit.

In Fig. 8.10, both an ideal and real LC-tanks are shown together with the necessary excitation so that they will start oscillating with their corresponding output signals. In this figure, a current pulse represents the necessary excitation. In real implementations, even small power resulting from the resistance noise is enough to provoke the oscillation of the output signal. Once oscillation has started, in the ideal scenario this state is indefinitely maintained. However, the oscillation in real LC-tanks is obviously dumped by internal losses in each cycle, unless the system is fed with extra power.

The different types of LC-tank oscillators differ basically in the way they generate negative resistance to compensate the losses in the tank. Eq. 8.11 illustrates the condition that must be met to maintain oscillation in the tank indefinitely, where G_p is the conductance equivalent to the tank and G_M is the negative conductance generated by the active circuit.

$$|G_M| > |G_P| \qquad \text{Eq. 8.11}$$

If the conductance generated by the active circuit is chosen to be exactly the same as the tank conductance, the condition required for steady state oscillation would be fulfilled, but initial transient effects could prevent this state being reached. In order to guarantee the start-up, it is recommended that the negative conductance is at least two times the positive conductance, but a higher ratio of three times is highly recommended. (Craninckx 1998).

8.3.2 Voltage Controlled Oscillator Design Constraints

Modern communication systems require ultra precise data management in high and wide frequency bands and, by hence, more precise noise treatment. Moreover, recent updates and new rehearsals of standards have highlighted the importance of mobile communications systems, which require by hence, low power consumptions in the whole receiver. Regarding to the oscillator design, the current

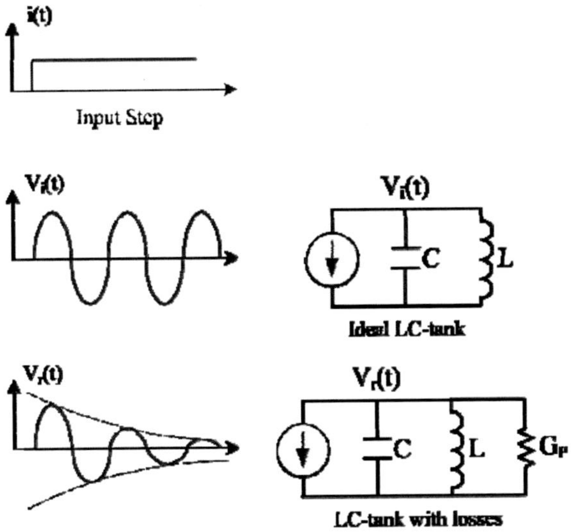

Fig. 8.10 Ideal and real LC-Tank transient output

consumption and the phase noise are critical concepts to take into account. Moreover, there is a direct dependence between these two parameters. Consequently, two highly extended phase noise models are presented with power consumption implications and a direct relationship with the passive elements of the circuit. But, first, the architectural decisions have to be adopted.

8.3.2.1 Topological Decisions

Once the general trends are set, the configuration has to be chosen. The following diagram (Fig. 8.11) shows all the existent architectures for integrated VCOs, and the highlighted path is the one under consideration in the examples of this chapter.

A tuned architecture must be chosen, because of the multichannel configuration that the VCO needs to be able to tune in for this standard, and for all the modern wideband standards. This main category is also advisable owing to the exigent phase noise requirement.

On the second level the LC structure is elected due to the full integration requirement for the nowadays-wireless communication devices working on the multi-GHz frequency bands.

The differential architectures provide a higher noise rejection with a good linearity and then are highly recommended.

On the final stage, any of the MOS structures may be used, albeit the CMOS permits to focus the design on the direct conversion zero-IF flicker noise mitigation and the low power, or the direct conversion to low-IF (Hajimiri and Lee 1999). Actually, the Negative Resistance Amplifier showed in Fig. 8.12 needs to overcome at least the permanent losses of the LC tank. However it would be recommendable to design it for twice these losses. This provides compensated

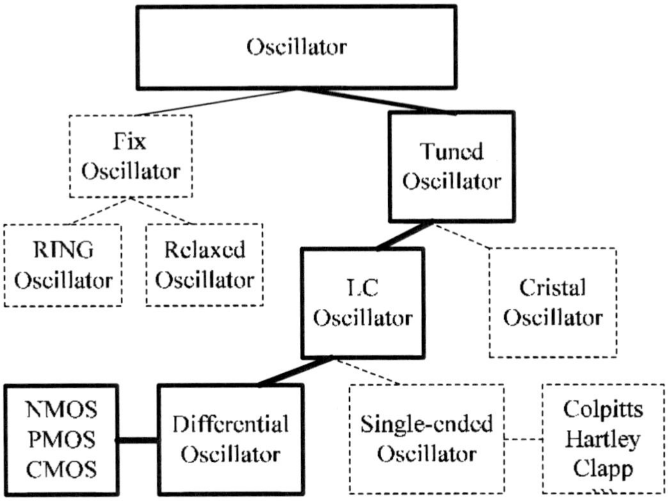

Fig. 8.11 VCO CMOS Topologies

constant amplitude for the signal, but supported by a higher current consumption. The CMOS topology minimizes it, in front of the only NMOS or PMOS cores construction due to the double compensation from the NMOS and the PMOS cross-coupled pairs working at the same time.

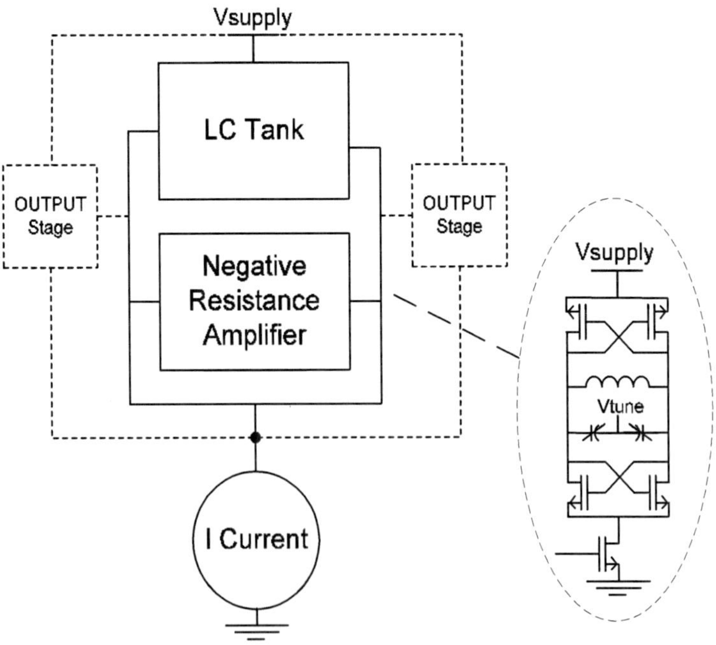

Fig. 8.12 VCO block diagram and core composition

As a drawback for this election, the PMOS and NMOS cooperation adds extra parasitic capacitance to the tank, since the transistors are both directly connected to the differential branches of the core. This reduces the tuning range, but it may be carefully controlled through the W/L ratio of both coupled pairs. As it has already been said, the over sizing contributes to the increasing of the power. This means that the cross-coupled NMOS and PMOS transistors, and even the current source transistors, should adequate their Widths and Lengths, reducing the ratio W/L, until the frequency band gap matches.

Solutions with NMOS cross-coupled solutions for the active part can be considered in certain cases, where the output power and phase noise requirements trade-off allows it.

The output stage is also pictured here, but it may be only considered depending on the necessity of connecting it to the following stage, and its input impedance.

8.3.2.2 Phase Noise and Current Consumption

Phase Noise is one of the most important parameters that characterize the functioning of an oscillator. This is why the specification of this figure of merit has a decisive influence on many design choices, e.g. the architecture of the oscillator, the tank circuit, and the active circuit. Therefore, in that subsection two different models commonly used are presented in order to evaluate the phase noise of a circuit during design stages, and their power consumption considerations.

In the temporal domain, the model of Hajimiri and Lee (see paregraph below), which considers an oscillator as a linear system but variable in time (LTV), stands out. This model does take into account the noise introduced by the active circuit and the conversion of flicker noise, but it does not provide an expression that allows for the estimation of phase noise.

On the other hand, in the frequency domain the model of Leeson stands out, and is based on the assumption that the oscillator is a linear system and temporally invariable (LTI). This model allows for a qualitative analysis of phase noise and enables us to obtain a mathematical expression for its estimation. The disadvantage of these models is that they do not take into account the conversion of flicker noise.

A - The Hajimiri and Lee Model

This approach is funded on theoretical predictions that adapt quite well to the characteristics of a real oscillator (Hajimiri and Lee 1998). This is the why this model is used more in simulators than in estimated calculations carried out in the first stage of designing the oscillator (Huang 2001).

This model assumes that the oscillator is a linear system variable in time (LTV) and studies the influence of total phase noise as much in active elements as in passive ones. Although the hypothesis of considering an oscillator as a linear system is not strictly true, in (Lee and Hajimiri 2000) it is shown that the effect of non-linearity of an oscillator on its phase noise can be neglected. This paper also provides a really interesting overview across the LC-tank and the ring oscillators.

According to this model, the phase noise of an oscillator can be calculated through the variation on the phase of the output signal when an impulse function is applied at the input. (Hajimiri 1998). From the impulse response of a LTV system it is possible to know the answer for any type of input and the phase noise from an offset frequency (Δf) from the carrier f_{off} results in Eq. 8.12:

$$L\{f_{off}\} = \frac{\Gamma_{rms}^2}{q_{\max}^2} \cdot \frac{\overline{i_n^2} / \Delta f}{16 \cdot \pi^2 \cdot f_{off}^2} \qquad \text{Eq. 8.11}$$

Where $q_{max}=C \cdot V_{max}$ and $\Gamma(\omega t)$ is a periodic function of period 2 called Impulse Sensitivity Function (ISF). When various sources of noise exist, $\overline{i_n^2} / \Delta f$ represents the total current in each node and is the sum in power of each one of the currents. This model of phase noise includes the noise at low frequency known as *flicker* noise in its analysis, pink noise or noise $1/f$.

It is important to point out that this model requires some excessively complex analytical calculations in order to be able to calculate the phase noise. As a consequence the deduction of an expression that provides the phase noise for a general case is not possible, but design recommendations are obtained from this model and it explains the procedure to follow in order to calculate this phase noise. The main conclusions of this model obtained in (Hajimiri 1998) are the following:

- As long as the oscillator operates in the limited current region, an area of operation in which the output amplitude is inferior to the supply voltage, an increase in the current from the oscillator provokes an improvement in its phase noise as a consequence of the increase in the amplitude of the output signal.
- The phase noise depends on the supply voltage.
- Considering the two previous conclusions it can be deduced that in order to obtain the best phase noise possible for a given consumption of power, the current from the oscillator must be increased and the supply voltage decreased in the same proportion. Logically this is only possible as long as the supply voltage is sufficiently high to bias the transistors and to ensure the oscillator functions in the limited current region.
- Lastly, it is important to point out the final recommendation of maintaining the maximum symmetry possible in the design of the oscillator in order to avoid the conversion of flicker noise as far as possible.

B - The Leeson Model

This model of phase noise (Leeson 1966) carries out the analysis in the frequency domain and assumes that the oscillator is a LTI system. It provides an estimation of phase noise at the output of an oscillator as a combination of a theoretical base, which presents certain simplifications, and empirical modifications introduced to fit the final spectrum to the real case. The expression that allows for the estimation of phase noise in dBc/Hz from the oscillator according to this model is presented in the following equation:

$$L\{\omega\} = 10\log\cdot\left\{\frac{2\cdot F\cdot k\cdot T}{P_{carrier}}\cdot\left[1+\left(\frac{f_0}{2\cdot Q\cdot \Delta f}\right)^2\right]\cdot\left(1+\frac{f_{1/f^3}}{|\Delta f|}\right)\right\} \quad \text{Eq. 8.12}$$

Where k is the Boltzman constant, T is the absolute temperature, Q is the quality factor of the LC-tank, f_0 is the frequency of oscillation, Δf the offset frequency with respect to the carrier, $P_{carrier}$ the power of the carrier, F an empirical factor that takes into account the increase in the noise density in the region $(1/\Delta f)^2$ (which typical value is close to 2) and $f_{1/f3}$ the frequency that limits the regions $(1/\Delta f)^2$ and $(1/\Delta f)^3$ (see Fig. 8.13).

In Eq. 8.12 it can be seen that, according to this model, the phase noise from the oscillator decreases when the power of the output increases together with the quality factor of the tank, whilst the offset frequency remains constant with respect to the carrier (Δf).

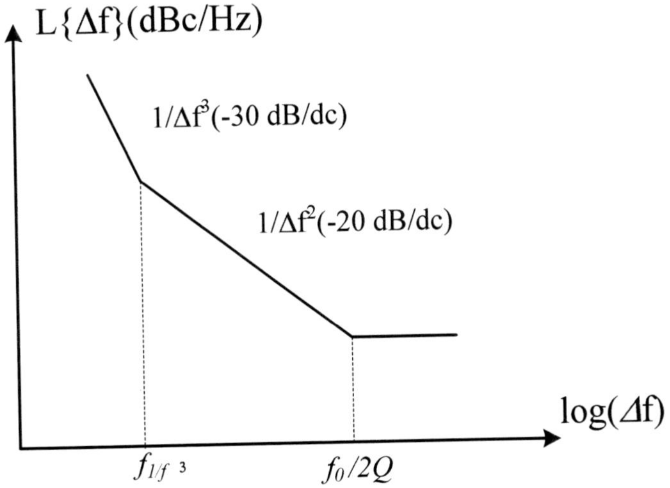

Fig. 8.13 Leeson Model graphical representation

The main drawback of this model is that it presents an empirical constant F and a value $f_{1/f3}$, both difficult to calculate in design stages. The former parameter (F) is a noise factor and completes the analysis considering the contribution of the tank losses, the current source and the differential pairs switching (Leeson 1966):

$$F = 2 + \frac{8\gamma R I_{source}}{\pi V_o} + \frac{8}{9}\gamma g_{mbias} R \qquad \text{Eq. 8.13}$$

Where I_{source} is the current source, γ the noise factor of a single MOS transistor (normally 2/3) and g_{mbias} is the transconductance of the differential pairs. Eq. 8.13 exhibits that the phase noise is inversely proportional to the square of the Quality factor and the Output Voltage of the tank. These two parameters are then focused for the design as they are related with the characteristics of the oscillator as well as with the current consumed. Furthermore this current appears in the noise factor numerator, which means that it adds directly its effect to the predicted phase noise expression. The first parameter under study, the voltage at the output, may be maximized but it is limited by the power supply down to a largest value of 2Vdd. But this nonlinear rise is supported by an augmentation of the current, and then a trade-off has to be settled. The optimum working point has been in (Leeson 1966) fixed at:

$$V_0 = \frac{4}{\pi} R I_{source} \qquad \text{Eq. 8.14}$$

This voltage makes the noise factor F constant (assuming g_{mbias}.R constant for this range of values) and consequently independent on the current consumed. Furthermore it represents a local point of interest for the L(Δf)-I curve: It is a minimum for the phase noise and it leads to the optimum current consumption calculated by the next formula.

$$I_{PN\min} = \frac{\pi V_{dd}}{4R} = \frac{\pi^2 V_{dd} f_0 C}{2Q} \qquad \text{Eq. 8.15}$$

This is the optimum current only considering the noise as the design objective. Nevertheless the power consumption is the main parameter of this study and then there is room for a further improvement in the generic optimum current. Actually, the threshold power needed by the following stage (deduced from Eq. 8.15) connected to the VCO should delimit the minimum current consumption. Consequently a trade-off between this minimum current and the optimized one obtained from the optimum phase noise expression has to be attained.

Another consequence here extracted is the dependence between the frequency and the current. It is directly related to the Quality Factor.

$$Q = \frac{f_c}{\Delta f} \qquad \text{Eq. 8.16}$$

This expression combined with Eq.8.15 means that a lower current provides a higher Q, and this brings a higher resonating frequency for a given bandwidth range. This might be a factor to consider in the nowadays wideband designs pushing the working bands up to, i.e., 11GHz for the UWB applications.

The last outcomes, highly related, are deduced from the expressions for the passive elements of the tank. For a given total quality factor (albeit it should be maximized for a better general performance) the inductance should be as high as possible, contrary to the capacitance value. However a higher L normally contributes a higher resistance to the tank that may degrades the whole quality factor. Therefore the recommended Q improvement methods may be followed to minimize the influence (Aguilera 2003): Geometrical optimization, Broken guard ring, Substrate shielding. As here clearly presented, the passive elements are crucial components of the Voltage Controlled Oscillators and have a big affection on the power consumption of the whole block.

8.4 High-Frequency Divider Design Constraints

This section presents the design of the high frequency blocks of frequency divider for RF integrated circuits. First, the basics on the frequency division are presented. Then, circuit design considerations are described: the Flip Flop topology has to be selected taking into account the power consumption considerations. With these latches, the architecture of the frequency divider and more specifically the High Frequency divider by 2 and the Dual Modulus Prescaler is faced. An implementation example is described in subsection 8.5.3.

8.4.1 Frequency Dividers Basic Implementation

In its most basic implementation the frequency dividers are basically made up of logic gates and flip-flops. They can be grouped in synchronous and asynchronous, depending on how the synchronization of these flip-flops is performed (Egan 2000). In the synchronous dividers, each flip-flop is triggered by the input signal of the divider (clock). On the other hand, in the asynchronous dividers the input signal of the dividerfeeds the first flip-flop, which triggers the second and so on. Therefore, the synchronous frequency dividers achieve a complete transition faster than the asynchronous ones.

Fig. 8.14a demonstrates a three stage asynchronous divider. Each stage consists of a divider by two, whose output is the input of the next stage. This makes the third stage asynchronous with respect to the input to the clock. If the outputs of the three stages are combined together with the input in an AND gate, the output signal in this gate will be synchronized with the clock (with a small delay caused by the gate). The main disadvantage of the asynchronous dividers is the accumulation of the jitter (time equivalent parameter of phase noise) from one stage to the other. This can be reduced placing a synchronization flip-flop at the end of the chain. In this case the jitter from the output is only that generated by the synchronizer (Levantino et al. 2004).

The synchronous dividers can be implemented using JK-flip-flops, as illustrated in the example given in Fig. 8.14b. Each stage changes almost simultaneously in the clock edges. This makes the last stage respond more quickly than in the case of the asynchronous divider shown in Fig. 8.14a. Special care needs to be taken when the intermediate outputs of each one of the stages are combined in any AND logic gate. This is due to the fact that undesired glitches may appear when any of the signals change more quickly or more slowly than the others.

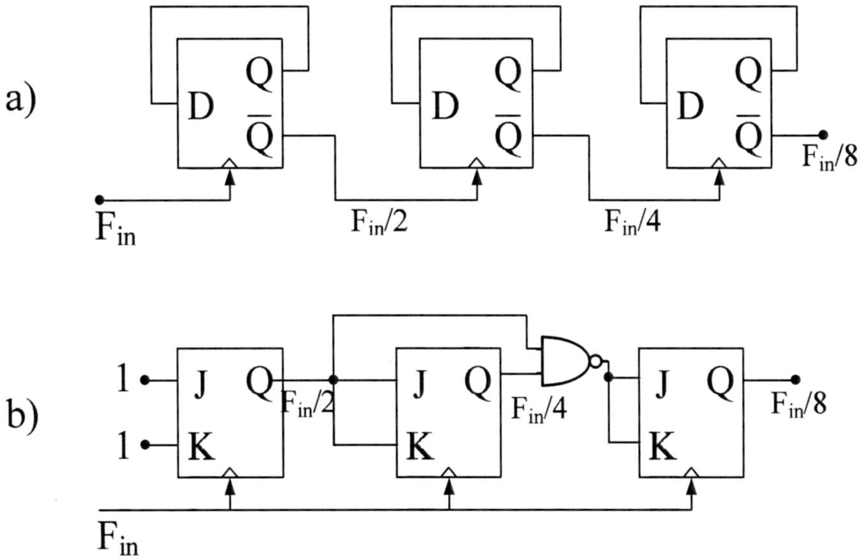

Fig. 8.14 Frequency dividers a) asynchronous b) synchronous

Although these basics could be used to achieve many different frequency reductions, the basic frequency divider may suffer from limitations for high frequency operations due to the presence of PMOS transistors and to their digital focused purpose. In those cases, the system must include more building blocks in order to accommodate the signal to the limitations of these dividers. That brings higher power consumption, and that is another reason why the high frequency operation always implies higher power consumption. In the next section we review the main architectures used for this operation and the design constraints, specifically designed for high frequency operation.

8.4.2 High Frequency Divider Architectures and Building Blocks

There are two main categories of high frequency dividers: fixed or tunable. As their name indicates, they differ in the type of division performed in each case. An adequate selection would depend on the channels' selection requirements for the application objective, as seen in the introductory section. Fig. 8.15 presents the basic diagram of both alternatives.

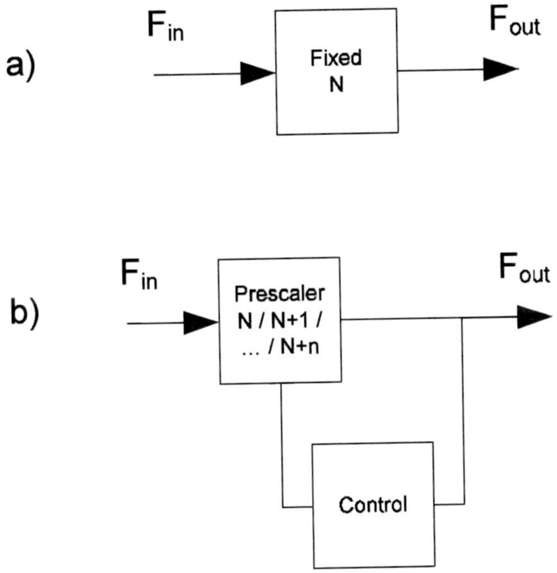

Fig. 8.15 Frequency dividers: a) fixed-N, b) Tunable

The Fixed-N divider is the simplest alternative as far as structure, dimension and current consumption are concerned but it is not valid for numerous applications that require implementation of a channel selector. This alternative, in Fig. 8.15a, is composed of fixed dividers, which progressively reduce the VCO frequency down to a suitable value to allow the comparison with the reference crystal signal.

With regard to tunable dividers, the most commonly used configuration is the architecture known as Pulse-swallow, which is composed of three blocks: prescaler, program counter and swallow counter, as illustrated in Fig. 8.16a. By contrast, the structure shown in Fig. 8.16b involves a more complex circuitry, including a Sigma-Delta modulator, but it also offers more flexibility to deal with different channel widths.

The difference between these two types of tunable dividers is the channel spacing that can be tuned by their control. In the first case in Fig. 8.16a the division ratio, limited to integer numbers in the first block, can only change as a combination of these integers. Whereas in the second case in Fig. 8.16b the shift between one integer to another in the first divider is not exclusively done at the end of a fixed counter count, producing fractional division ratios, and by hence a wider tuning capability.

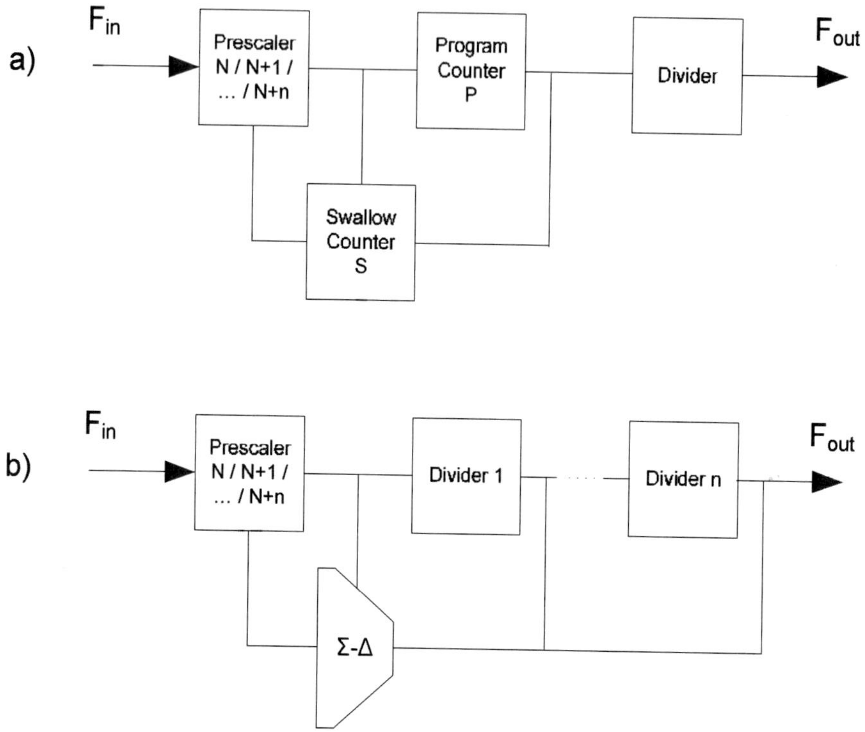

Fig. 8.16 Tunable Frequency dividers a) Pulse Swallow structure b) Sigma-Delta structure

It is interesting to note that all of these circuits are based on the same basic cell, called latch, and for this reason we will dedicate a separate subsection to the study of this block and to analyse their power consumption considerations.

Focusing our attention on the prescalers, the most commonly used for both types of tuneable dividers are the dual modulus prescalers (DMP), which can perform the division by two preselected values. These values can be selected through a specific control input.

Once the input frequency has been divided by the DMP, the rest of the blocks usually operate at significantly lower frequencies. For this reason the program counter, swallow counter and subsequent dividers may be designed using digital libraries.

Finally, it is interesting to note that in many high frequency PLLs the first block of the frequency divider is a divider-by-2. With this strategy the specifications of the rest of the building blocks are conveniently relaxed. The resulting architecture for the PLL would then be the one shown in Fig. 8.17.

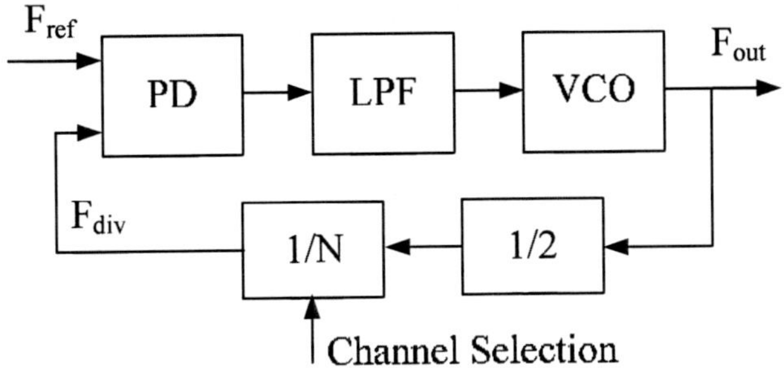

Fig. 8.17 PLL with a divide-by-2 circuit as the first module of the frequency divider

8.4.2.1 High Frequency Divider-by-Two. Flip Flop Design for Low Power Consumption

The circuits of division by two, also known as DTCs (Divide-by-TwoCircuits), are widely used to generate quadrature outputs. However, because of the fact that these circuits can reach speeds higher than dividers with other factors of division, the DTCs can also be employed in a PLL in order to reduce the VCO frequency down to an appropriate value. This ability means the block is the popular choice in these kinds of systems.

The different architectures found to design DTCs at frequencies in the order of GHz can be classified into static latches in master/slave configuration, dynamic latches and Miller divider latches. Outlined below is the main characteristic of each:

Static latches in master/slave: The first architecture, known also as JohnsonCounter, is illustrated in It consists of two latches connected in master/slave configuration. Each flip-flop is triggered by two complementary clock signals, CLK and $\overline{CLK}$. The two flip-flops work periodically and alternatively between two modes. When the input signal CLK is at a low level, one of the latches is in sensing mode (it receives the data at its input D and it copies them into its output Q), while the other is in latch mode (it keeps the previous output). When the input signal CLK goes to high level, the flip-flops exchange their modes of operation. This mechanism means that the frequency from the output signal (Q and $\bar{Q}$) can be half that of the input (CLK and $\overline{CLK}$).

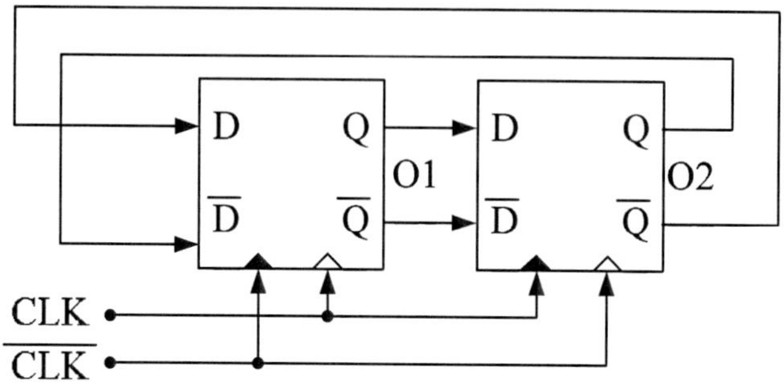

Fig. 8.18 Jonson Counter block diagram with DTCs

This architecture provides perfect quadrature in O1 and O2 only if the signals CLK and $\overline{CLK}$ are exactly complementary, i.e. the mismatch between both latches is practically nonexistent. In real conditions the deviations can be translated in differences in phase of up to 5 degrees. Furthermore, if CLK and $\overline{CLK}$ are not exactly differential, additional imbalances of phase will appear.

Diverse topologies have been found for the implementation of each one of the latches in this architecture, optimised to achieve maximum operating frequency and minimum power consumption, being the most commonly used the configurations proposed by Razavi, Wang and the SCL type (SourceCoupleLogic).

Dynamic latches: Unlike statics, the implementation of dynamic latches is not based on a bi-stable circuit. The parasitic capacity between the nodes acts as a storage element. The dynamic architecture most commonly used is that which employs a TSPC (TrueSingle-PhaseClocking) latch.

Miller divider: This high speed method for dividing by two, originally proposed by Miller and explained in generic books as, i.e. (Razavi 1998), is illustrated in Fig. 8.19.

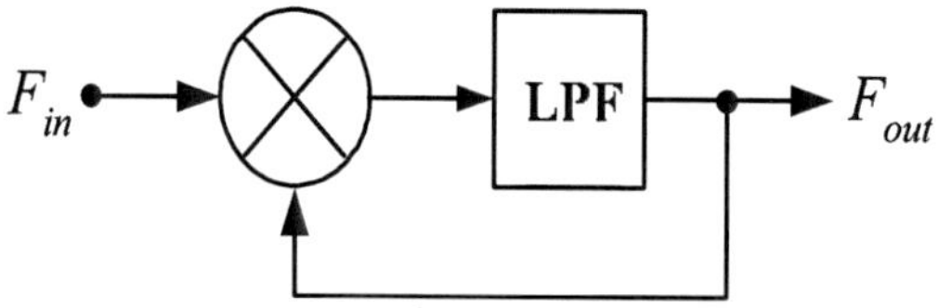

Fig. 8.19 Miller Divider

Using a mixer and a low-pass filter, this configuration operates in the following way. After carrying out the mixing process, the output from mixer components at frequencies of $F_{in}+F_{out}$ and $F_{in}-F_{out}$ are generated. If the first is suppressed by the filter but the last one is not, then $F_{in}-F_{out}=F_{out}$, and therefore $F_{out}=F_{in}/2$. The simplicity of the loop of transmission allows this topology to operate at speeds higher than half of the f_T of the transistors, regularly achieving the highest speeds out of all divider configurations. However, the phase noise introduced is too high for practical RF applications, since nor the mixer, neither the LPF are low noise oriented. For this reason, this configuration is not commonly used in high frequency PLLs.

In the next paragraphs, the most common configurations for RF applications are explained in greater detail.

Razavi

The schematic circuit of this latch in master/slave configuration is shown in (Razavi et al. 1995). In high-speed dividers with master/slave topology the slave latch is usually designed as the master dual so that both can be driven by the same clock signal. However, the duality requires one of the latches to incorporate PMOS transistors in the path of the signal, reducing the maximum speed of the divider. In order to avoid this, identical latches that are driven by the complementary clocks CLK and $\overline{CLK}$ are used.

In Fig. 8.20 each latch consists of two transistors working in sense mode (M1 and M2 in the master and M7 and M8 in the slave), a regenerative loop (M3 and M4 in the master and M9 and M10 in the slave) and two transistors acting in pull-up mode (M5 and M6 in the master and, M11 and M12 in the slave). When the clock is at high level, M5 and M6 are cut-off and the master is in sensing state, while M11 and M12 are activated and the slave is in latch mode. When the clock changes to low level, the opposite process occurs.

The most important advantage of this topology is its speed, given that it does not use stacked PMOS transistors and the current flows just through two gates per cycle. On the other hand, it presents the disadvantage of consuming static energy (it consumes during the whole cycle) and requires a differential clock input signal (Perrott 2003).

Wang

A variation of the Razavi topology has been proposed by Wang (Wang 2000), whose diagram is presented in Fig. 8.21. The differences in this configuration with respect to that of Razavi is the inclusion of NMOS clock transistors in each one of the two latches (M13 in the master and M14 in the slave) and the different layout of the PMOS clock transistors.

Despite the similarities between both topologies, the working principle of them is slightly different. In the Wang architecture the operation mode of a flip-flop (sensing or latching) is controlled by a couple of switches M13 and M14 rather than by the PMOS transistors. Furthermore, the resistance from the PMOS transistors is low in the sensing mode and high in the latching mode, something

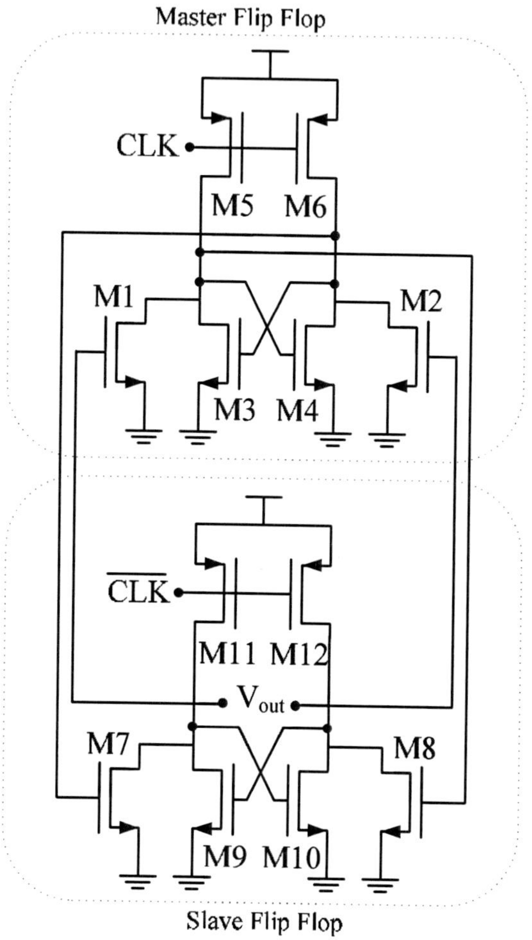

Fig. 8.20 Razavi topology for the DTC

that makes the time constant smaller and the signal that attacks the other flip-flop higher, in contrast to the Razavi topology.

With this solution, the circuit can operate at higher frequencies, without increasing the power consumption. For example, comparing (Razavi et al. .1995) with (Wang 2000), the maximum operating frequency of the Wang architecture is superior to 18 GHz with a supply voltage of 1.8 V and CMOS technology 0.25 μm. On the contrary, the Razavi topology reaches a working frequency of around 9 GHz for the same supply voltage and a CMOS technology 0.1 μm. On the other hand, the consumption of power in the last one is higher for the same supply voltage. For example, if the supply voltage is 1.8 V, the Razavi topology consumes around 10 mW, while that of Wang 3.6 mW.

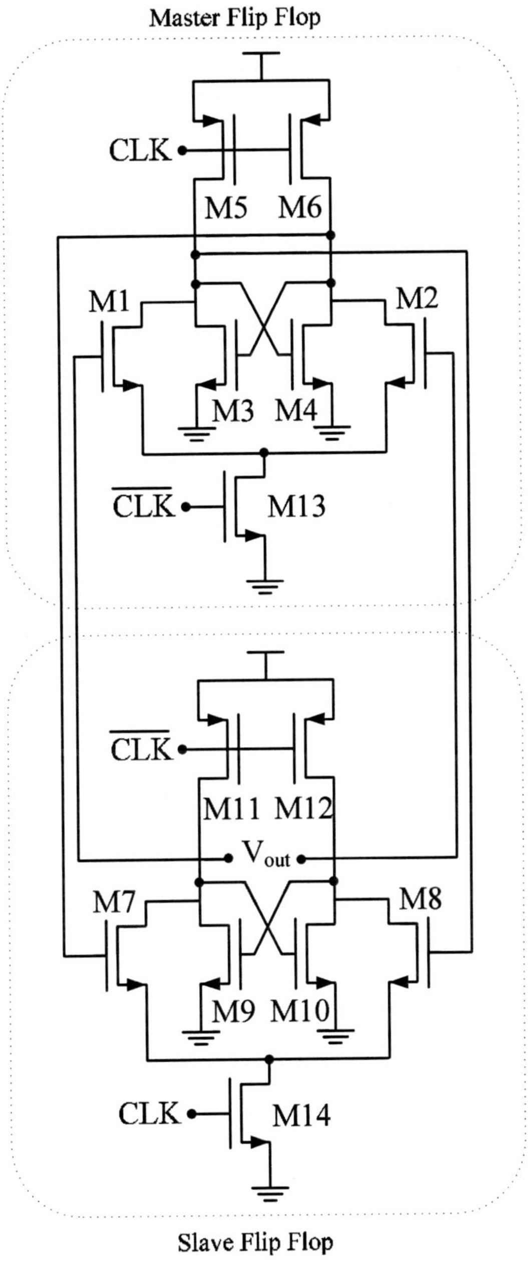

Fig. 8.21 Wang topology for the DTC

SCL

The last topology found to implement each one of the latches in master/slave configuration is the SCL (SourceCoupleLogic), in which each of the flip-flops are present in the circuital diagram in Fig. 8.22 (Wohlmuth et al. 2002).

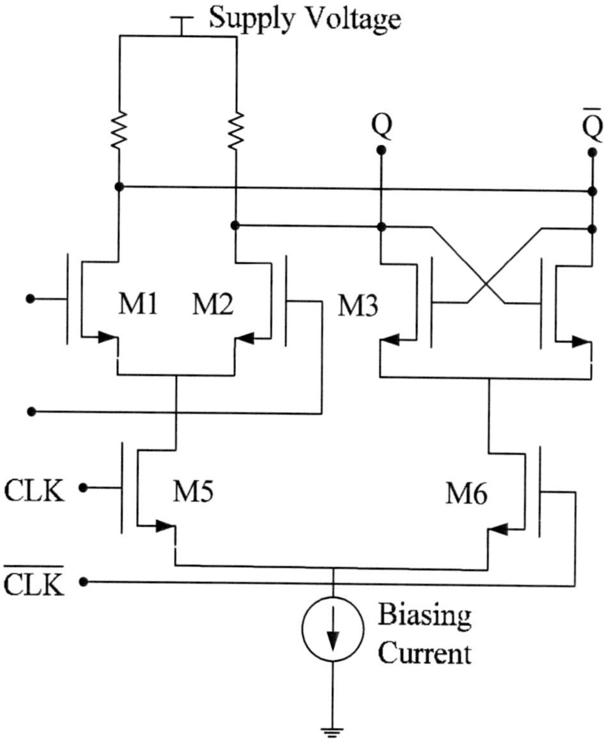

Fig. 8.22 SCL Flip-Flop topology

This topology eliminates the PMOS clock transistors and uses resistances as load. Due to the total absence of PMOS transistors and to the fact that the signal only circulates through two gates per cycle this structure is faster than the two previous ones. In addition, it functions correctly for dynamic ranges of the input clock smaller than in the previous designs. (Perrott 2003). Lastly, this topology is the most suitable to work at frequencies of various GHz. A careful selection of transistor size and type allows for a reasonable equilibrium between speed and power consumption at operating frequencies of GHz (Razavi 1998), and provides robustness in the phase noise coming from ground and supply tracks.

On the other hand, it also presents some disadvantages compared to the previous two architectures. As well as the static power consumption and the need for differential clock signals, the occupied silicon area is higher and requires the implementation of current sources (Perrott 2003).

In (Wohlmuth et al. 2002) the implementation of a DTC with SCL topology in a CMOS 0.12 μm technology is described. For a supply voltage of 1.5 V an operating frequency of 27 GHz is reached, with a power consumption of 45 mW. It can be seen how the power consumption is superior to that of the Wang and Razavi topologies for similar supply voltages. However, the operating frequency is very superior in this SCL topology.

TSPC

Unlike the static architectures, this topology does not consume static power, dissipating power only in the clock transitions. In addition to this advantage, it can be implemented with less chip area consumption and it does not require differential clock signals. On the contrary, it is a slower topology than the static ones due fundamentally to the use of stacked PMOS transistors and that the signal circulates through three gates per cycle (Perrott 2003). Nevertheless, that topology can be used in applications in the 5GHz U-NII band (Adin 2007), and by hence is considered optimum from the low power point of view.

This structure basically differs from the classical Common Mode Latch (CML) in the absence of current source. There are two solutions based on this TSPC structure:

a) The classical TSPC FF is a dynamic circuit that only consumes power during switching. The functional schematic of this circuit is shown in Fig. 8.23a. Since the publication of this technique (Yuan and Svensson 1989), the downscaling of the CMOS technology has surfaced its frequency capabilities with minimal power consumption (Navarro Soares1999), (Pellerano et al. 2003).

b) The Extended TSPC (E-TSPC) FF has been aimed to improve the frequency performance in comparison with the classical one. The basic schematic is depicted in Fig. 8.23b. The number of transistors of this FF decreases from 9 to 6 in this novel E-TSPC structure; hence, the transition delay is lowered. The operating frequency of any digital device is restricted by the limitation delays of the high-to-low and the low-to-high transitions. Furthermore, the power dissipation due to switching is directly dependent on the frequency of this delay. Nevertheless, this progress resulted in the increasing of the total power consumption. The E-TSPC FF has another addend in the total power dissipation that is not present in the TSPC structure: E-TSPC has short circuit power dissipation when the PMOS and the NMOS transistors were turned on simultaneously. This is a residual effect produced in a quarter of cycle of a simple division by 2. In average, the total power dissipation is increased in 200% (Yu et al. 2006).

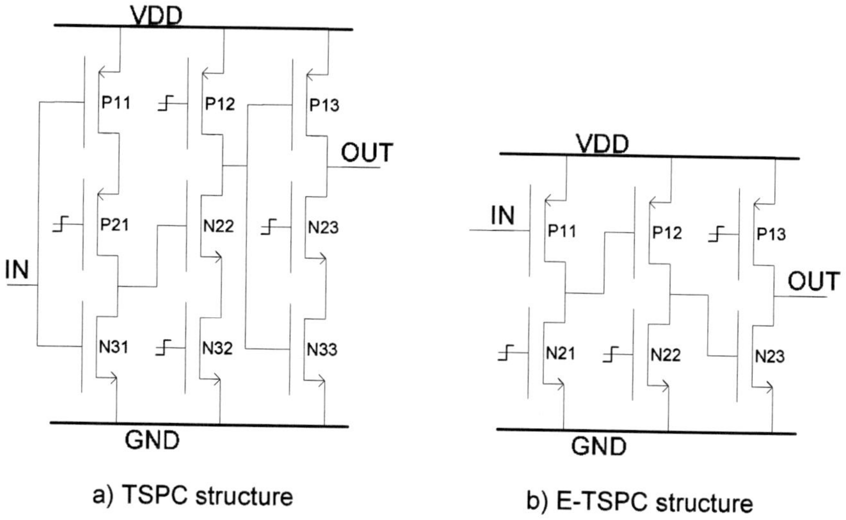

Fig. 8.23 TSPC and Enhanced TSPC D Flip Flop structures

Therefore, the TSPC structure will be used when the system behavior is not affected by the transition delay.

8.4.2.2 Dual-Modulus Prescaler

As already explained in the introduction of this section, most tunable dividers incorporate a dual-modulusprescaler. This type of high frequency divider divides the input signal by a rate of division M or M+1 depending on an input control. The values of the division ratio can be very different depending on the requirements of each application.

The prescalers mainly consist of flip-flops and high operating frequency logic gates. The configurations used for implementing high frequency latches described in the previous section can be used to implement the flip-flops in this type of prescaler.

As there is no need to present a classification of dual-modulusprescalers depending on their division ratio, the following describes the most basic and commonly used type, the prescaler that divides by 2 or 3, whose circuit diagram is illustrated in Fig. 8.24b.

In order to explain the working principle of this circuit, in the following paragraphs we are going to describe the 2/3 divider. Prior to this description, the truth table of a divider by 3 is presented in Table 8.1, while Fig. 8.24a shows its block diagram. This divider employs two type D flip-flops in master/slave configuration with a logic gate AND in order to create only the three states shown in Table 8.1

Table 8.1 Logic state for the divider by 3 of Fig. 8.24a

Q	0	1	1
$\bar{Q}$	1	0	1

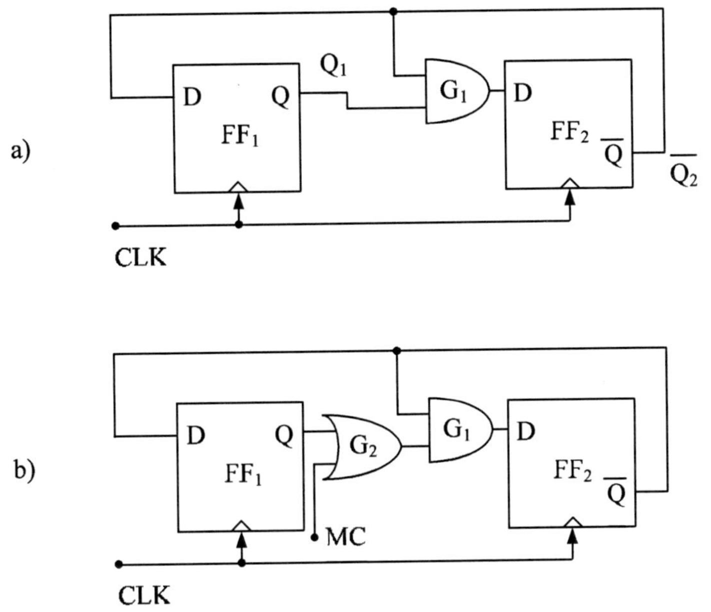

Fig. 8.24 a) Divider by 3 b) Divider by 2/3

To convert the topology in Fig. 8.24a into a 2/3 prescaler introducing a logic gate OR between the first flip-flop and the gate AND is sufficient. In this way, when MC is at a high level, the prescaler will divide by 2 and when it is at a low level it will divide by 3. Due to the existence of logic gates between the two flip-flops this type of divider is slower than the high frequency dividers by two presented in the previous section.

8.5 Low-Power Design Examples

This section describes in detail the design of two different low-power LNA examples, regarding its circuit design considerations, layout implementation, the test setup used for the measurement of their main parameters, the characterization results, and finally a short discussion on such results.

- The first example is a low-power shunt-series feedback-based wideband LNA for the DVB-T and DVB-H standards (DVB-T_EN300744), (DVB-H_EN302304).

- The second example shows another low-power LNA for higher frequency applications in the 5GHz U-NII frequency band, such as WLAN (IEEE99), (IEEE03).

8.5.1 Example 1: Wideband VCO for DVB-H

The architecture chosen for the VCO implementation is the cross-coupled NMOS with LC-tank, due to its good performance regarding phase noise. The simplified schematic of the VCO is shown in Fig. 8.25.

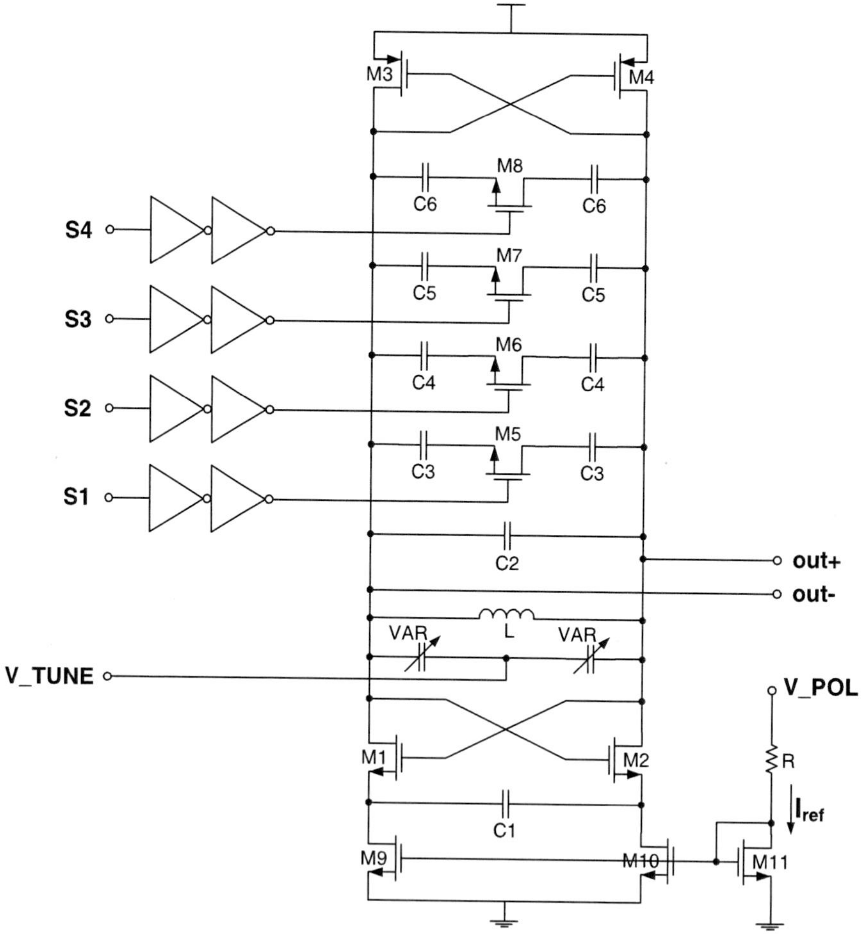

Fig. 8.25 VCO core schematic of the DVB-H VCO example

The following sub-subsections describe the VCO circuit design with respect to its main building elements, that is, cross-coupled CMOS pairs, LC-tank, active devices and tail device respectively.

8.5.1.1 Cross Coupled CMOS Pairs

The transconductance stage is composed by devices M1 through M4. Such architecture is called cross-coupled CMOS as it is composed of complementary devices. It exploits the current reuse technique, previously presented in that book, therefore, for the same current consumption the transconductance is doubled, and thus the differential voltage swing is larger, resulting in reduced phase noise. Another important benefit in this architecture is that 1/f noise is minimized.

Minimum channel length devices have been used to reduce the parasitic temperature dependent capacitances. The width of these transistors represents a trade-off between power consumption, phase noise and tuning range. Larger devices present higher transconductances and hence better phase noise performance, including improved 1/f noise characteristic. However, the achievable tuning range is reduced, since the parasitic drain capacitances are directly added to the LC-tank, reducing the bandwidth. Devices M1 through M4 are 100μm width.

8.5.1.2 LC-Tank

Before starting the design considerations of the LC-tank, a first assumption regarding to the VCO output frequency must be set. The LC-tank has been designed to work at twice the frequency required by the DVB-H standard specification. The main reasons to do so are:

- The tuning range relative to the carrier frequency (in %) is reduced.
- Both inductance and capacitance reactances are smaller. Thus, higher quality factors can be achieved in CMOS technologies with smaller devices.
- Although the phase noise contributed by the active devices is higher at higher frequencies, it is compensated by the increase of the quality factor of the passive devices (specially the inductor), as can be deducted from Eq. 8.2.
- The quadrature output is obtained directly from the outputs of the divide-by-two circuit that follows the VCO.
- In zero-IF receivers, the self-mixing and LO re-radiation problems are minimized doubling the frequency of the VCO.

Therefore, the VCO has been designed to provide an output frequency ranging from 948MHz to 1.716GHz. The LC-tank provides considerable attenuation at frequencies other than the oscillation frequency, thereby giving better phase noise performance. As all the elements of the oscillator, the LC-tank components are full differential, in order to reduce the sensitivity to power supply variations and substrate coupling. It is very important to design the LC-tank in such a way that its quality factor is maximized, since it directly affects the phase noise performance.

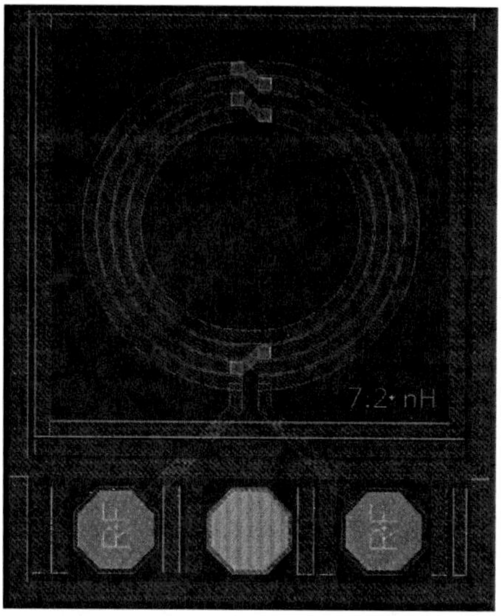

Fig. 8.26 Differential inductor layout for DVB-H VCO example

The Q of the inductor is the most critical one, and is directly related to the series parasitic resistance. Larger inductance values improve the phase noise performance, but in real implementations the series resistance is also higher, so the dimensions of the inductor must be carefully selected. The top metal has been used for the inductor implementation, due to its lower resistivity and lower parasitic capacitance to the substrate. For a differential inductance of 7.2nH, the inductor has been designed using 4 turns, an external radius of 180μm, 15μm of track width and 2μm for the track spacing. Its layout is shown inFig. 8.26.

The varactor has been directly taken from the passive devices library provided by the foundry. The MOS varactor has been implemented as 264μm width (4 rows and 10 columns), with a Cmax of 837.41fF.

8.5.1.3 Array of Switched Capacitors

As has been previously outlined in this example, to cover the wide tuning range of nearly 800MHz with a single varactor is a difficult issue. In addition, the required VCO gain would be large, amplifying any noise coupling to the control node and hence deteriorating the phase noise performance. Therefore, an array of binary weighted switched capacitors has been implemented to achieve the required tuning range.

The most important limitation to the application of this technique has to do with the degradation in the Q of the tank due to the ON-resistance of the switches. Hence, the use of rather large switch devices in the coarse-tuning capacitive array is needed. This increases the switch off-capacitance; requiring a smaller inductor for a given maximum frequency and thus worsening the sensitivity to layout parasitics. All the switches have been designed to have a width of 225μm and the minimum length for a better switching efficiency and 1/f noise performance.

For a given tuning range, the optimum number of coarse-tuning bits maximizes the tank inductance and minimizes the varactor size. A large number of coarse-tuning bits result in large off-capacitance and thus lower inductor value. On the other hand, fewer than an optimum number of tuning bits result in large varactor and subsequently lower inductor value. A 4-bit digital word has been used in this design to coarse-tune the VCO (S1 through S4 in Fig. 8.25). The unit capacitance has been designed to be 460fF.

The use of two identical tuning circuits enables switched tuning of a differential oscillator (Sjoland and Sjoblom 2002). When a branch is turned ON it will contain two ON-state resistances in series going from one side of the differential circuit (via GND) to the other. This limits the achievable quality factor. If the transistor retains the dimensions, the OFF capacitance is the same, as the transistor, which is assumed to be symmetric, has equal source and drain capacitances. The quality factor on the other hand is doubled, since just one device ON resistance is connected in series with the capacitors.

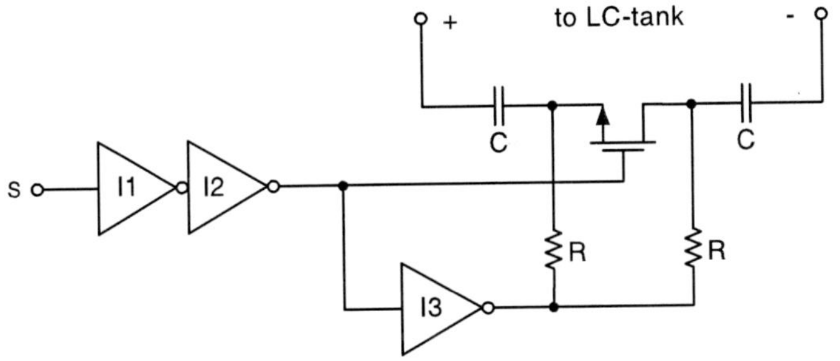

Fig. 8.27 Switched-tuning circuit implementation for DVB-H VCO example

The inverters do not consume any static power, and they occupy a negligible amount of silicon area. Inverter I3 together with the resistors makes sure that the transistor is OFF at all times in the OFF state, and gets the maximum gate to source (and drain) voltage in the ON state. On the other hand, inverters I1 and I2 low-pass filter the noise coupled to the control lines, contributing to a better phase noise performance.

8.5.1.4 Tail Current Source

The tail current source of the circuit in Fig. 8.25 is created by the tail devices M9 and M10. Varying the width of these transistors negligibly affects the phase noise, whereas increasing the length of the devices reduces the flicker noise and hence their contribution to the oscillator's phase noise. Thus, they have been implemented as 80μm width with a channel length of 0.6μm. Each of them drives a DC current of 2.15mA, set by the voltage V_TUNE. Therefore, the VCO core current consumption is 4.3mA.

8.5.1.5 Layout Considerations

One important issue of VCO layout design is the metal width. The resonating tank causes the current in the tank to be Q times larger. Hence, the metal tracks connecting the tank elements need to be wide enough to withstand the large currents.

Another consideration with any differential circuit is the symmetry between the two halves of the circuit. Any asymmetry in the two sides of the oscillator will cause an unbalance in the tank causing the center frequency to change. Since the resistance of the metal tracks degrades the quality factor of the tank, they must be as short as possible. A view of the layout of the core of the VCO is shown in Fig. 8.28.

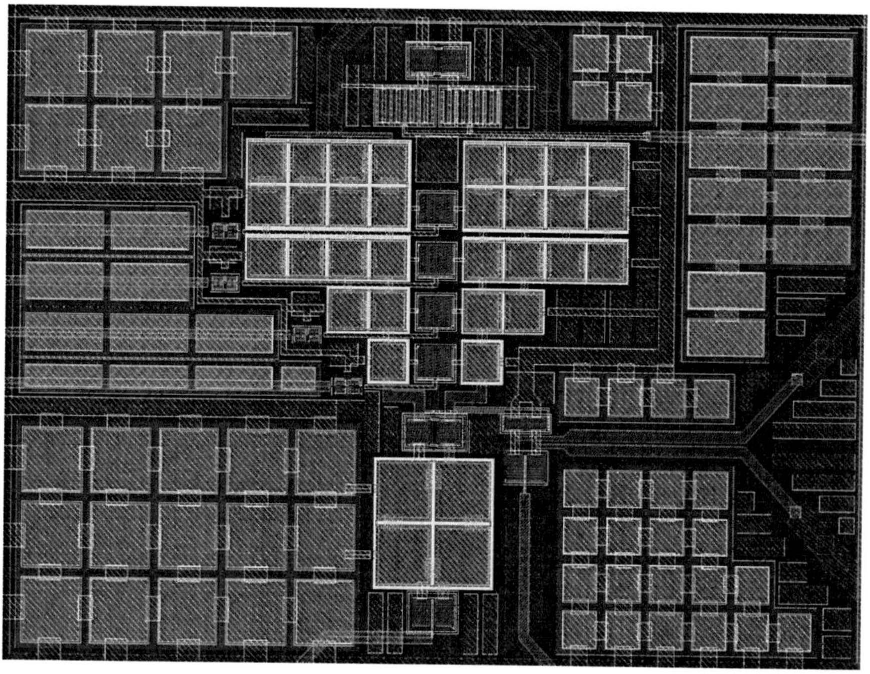

Fig. 8.28 VCO core layout for DVB-H VCO example

Resistance has been further reduced by using the top metal layer due to its lower resistivity. The gates of the cross-coupled transistors have been connected on both ends with metal tracks, reducing the gate resistance, thereby improving the tank quality factor. All devices incorporate guard rings to better isolate against substrate coupling noise. Extra substrate contacts have been placed strategically close to the transistors' gate.

The VCO complete layout is showed in Fig. 8.29. As many capacitances as possible, have been placed from V_{DD} to ground to reduce the influence of the noise coming from the power supply lines. In a similar manner, the die area has been filled out with substrate contacts to ground, for an optimum performance against substrate noise coupling. All the DC control lines incorporate de-coupling shunt capacitances to low-pass filter any noise coming from these lines.

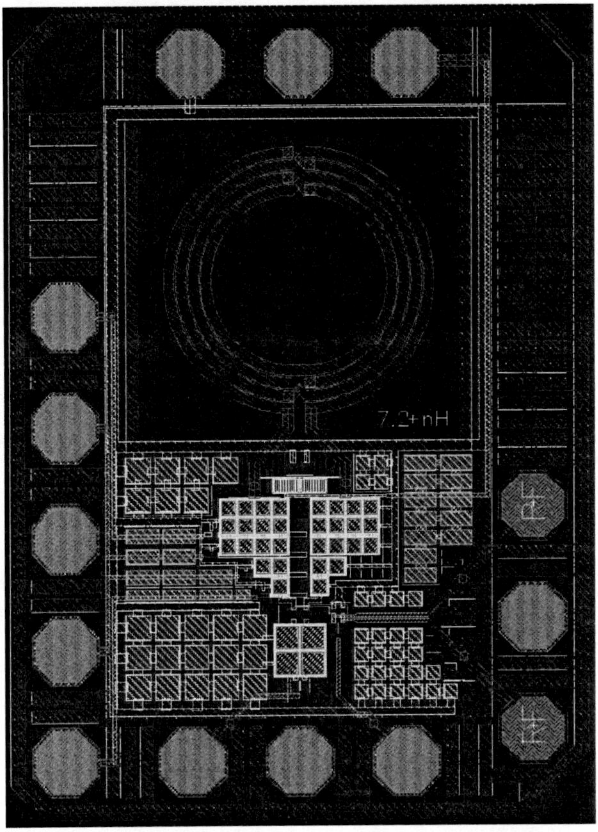

Fig. 8.29 VCO complete layout for DVB-H VCO example

8.5.1.6 Test Setup and Characterisation Results

The VCO has been measured on-wafer. Its die photograph is shown in Fig. 8.30. A detail view of the VCO core is displayed in Fig. 8.31.

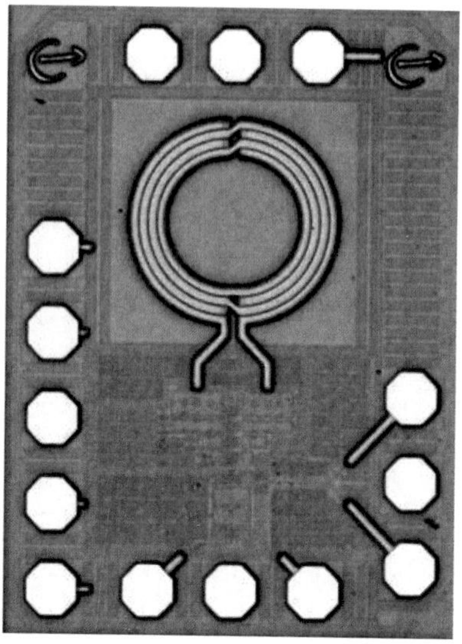

Fig. 8.30 VCO die photograph for DVB-H VCO example

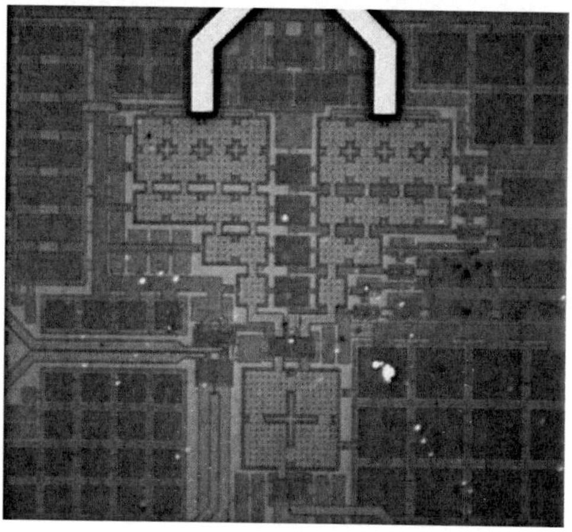

Fig. 8.31 VCO core detail microphotograph for DVB-H VCO example

The measurement setup is depicted in Fig. 8.32. Due to the high sensitivity of the VCO to the noise coupled through the digital control lines (S1 through S4), they have been low-pass filtered by means of an additional PCB. V_POL1 and VPOL_2 DC sources set the proper current values for the VCO core and output buffer respectively. The latter has been added only for testing purposes.

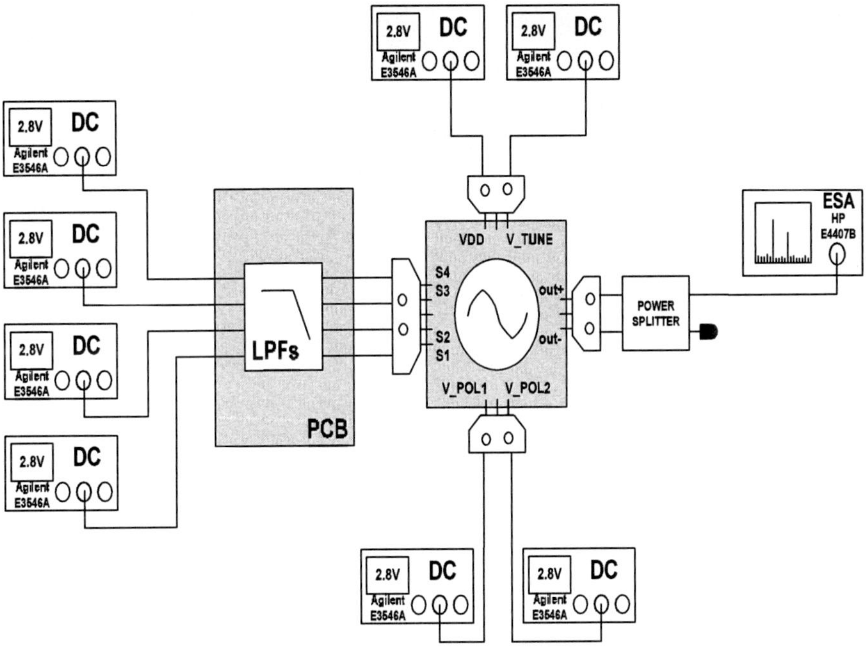

Fig. 8.32 VCO measurement setup for DVB-H VCO example

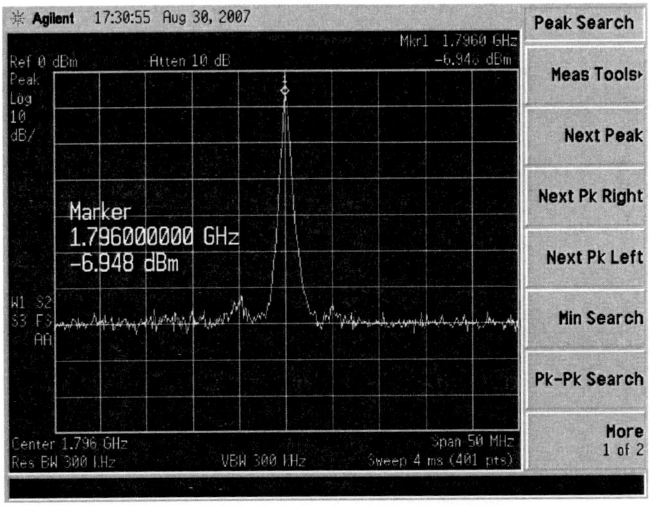

Fig. 8.33 VCO output signal screenshot for DVB-H VCO example

The output signal of the VCO is directly measured in the spectrum analyzer. A screenshot of the upper frequency signal is shown in Fig. 8.33. Fig. 8.34 and Fig. 8.35 show the tuning range and the output power respectively, as a function of the tuning voltage (V_TUNE).

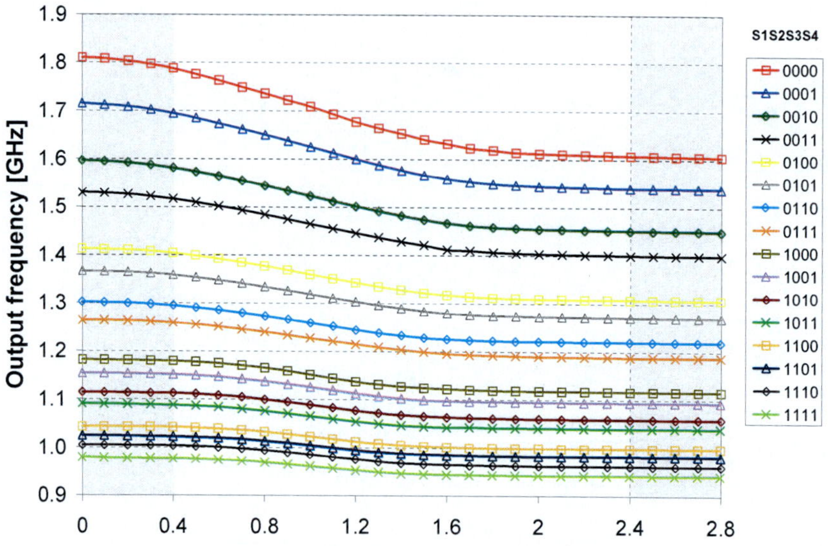

Fig. 8.34 VCO tuning range vs. tuning voltage (V) for DVB-H VCO example

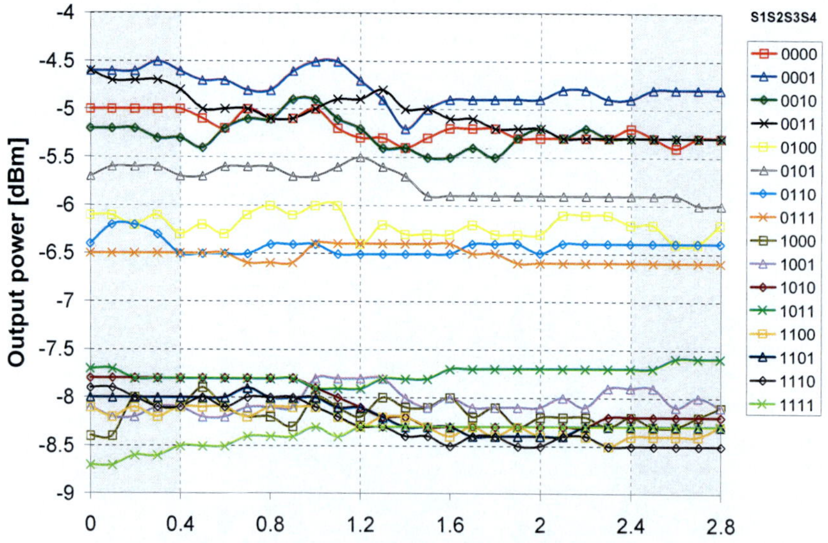

Fig. 8.35 VCO output power vs. tuning voltage (V) for DVB-H VCO example

Table 8.2 lists the different tuning ranges and average output powers depending on the value of the control word that selects each of the sub-bands.

Table 8.2 Tuning range and average output power for all the sub-bands

S1S2S3S4	f_{min} [MHz]	f_{max} [MHz]	$P_{out,av}$ [dBm]
0000	1605.0	1810.0	-5.2
0001	1537.5	1715.0	-4.8
0010	1449.4	1596.5	-5.3
0011	1398.8	1529.5	-5.1
0100	1306.8	1413.4	-6.2
0101	1269.3	1366.2	-5.8
0110	1217.6	1302.3	-6.4
0111	1186.8	1265.0	-6.5
1000	1115.7	1181.6	-8.2
1001	1092.5	1154.2	-8
1010	1058.3	1114.2	-8.1
1011	1038.7	1091.1	-7.7
1100	997.0	1043.5	-8.3
1101	980.5	1024.3	-8.2
1110	960.5	1004.3	-8.3
1111	940.7	979.8	-8.4

With reference to phase noise, the procedure for a rough estimation at 1MHz offset is as follows: the power of the signal at 1MHz offset from the peak is 80dB lower than the peak power. Due to the fact that the resolution bandwidth of the spectrum analyzer is set to 10kHz, the phase noise can be expressed as -80dBc/10KHz = -120dBc/Hz. Considering a total signal loss of 2.5dB and that the peak power is 3dB lower than its nominal value the phase noise has been estimated to be PN = -125.5dBc/Hz. Nevertheless, this is a worst-case estimation. In addition, the good agreement between the measurements and post-layout simulations predicts the phase noise to fulfil the design requirements.

Finally, Table 8.3 summarizes the VCO performance, in terms of the measured results vs. the initial requirements.

Table 8.3 VCO measurement results summary

Parameter	Unit	Specification	Measurement
Tuning range	MHz	948 - 1716	940 - 1810
Output power	dBm	> -10	> -8.4
Phase noise @ 1MHz	dBc/Hz	-126.5	-125.5
Power consumption	mW	min	12

The VCO fulfils all the requirements stated with a current consumption of 4.3mA. Its performance can be summarized as follows:

- The tuning range extends from 940MHz up to 1810MHz among 16 sub-bands.
- The output power is higher than -8.4dBm (up to -5.2dBm for the upper sub-band)
- A phase noise of -125dBc/Hz measured at 1MHz offset.

Regarding to the frequency response of the VCO, the whole band has been covered among 16 sub-bands. However, the upper bound (1810MHz) is 94MHz higher than the requirement (1716MHz), whereas, the lower bound (940MHz) is only 8MHz lower than the specification (948MHz). This issue can be corrected by re-designing the varactors of the LC-tank. A wider varactor would correct the frequency deviation, translating the frequency response of all the sub-bands towards lower frequencies.

In the same way, depicted in Fig. 8.35 is the output power of the VCO as a function of the tuning voltage. The output power is nearly flat for all the sub-bands (0.7dB max. variation). In any case, the measured output power (referred to 50Ω) is higher than the specification (-10dBm).

Finally, the rough estimation of the phase noise (measured at 1MHz) gives as a result -125.5dBc/Hz, which is 1dB lower than the specification. However, post-layout simulations predicted a phase noise of -128dBc/Hz in a worst-case scenario (considering all the sub-bands). Considering the good agreement between post-layout simulations and phase noise measurements in similar implementations carried out by the design group (Mendizábal 2006), it can be stated that the phase noise requirement is fulfilled by the VCO.

8.5.1.7 Discussion

Table 8.4 presents a comparison of the most representative broadband VCOs collected in the bibliography, in the terms of their most relevant parameters:

- Tuning range ($f_{min} - f_{max}$)
- Phase noise extrapolated to 1MHz offset
- Power consumption
- Fabrication Technology

In this case, a comparison between the VCO described in this example and other switched-capacitance wideband VCOs has been carried out, due to the lack of reported results from specific DVB-H VCOs.

It is difficult to compare the devices in terms of phase noise, as the phase noise reported in this example is a rough estimation. However, all the VCOs with lower phase noise present a tuning range also lower (relative to the output frequency). In addition, post-layout simulations predict a phase noise performance only surpassed by (Dawn et al. 2002).

Regarding power consumption, this work is only exceeded by (Berny et al. 2003) (lower tuning range and worse phase noise) and (Kampe et al. 2005), which has been implemented in a 0.18μm CMOS technology. This way, although the current consumption is higher than the reported in this work, the power consumption is lower as the supply voltage has been scaled down.

Table 8.4 State-of-the-art broadband-low phase noise VCOs

[REF]	tuning range (GHz)	phase noise @ 1MHz (dBc/Hz)	power consumption (mW)	technology (μm)
(Manetakis et al 2004)	2.8-4.55	-128	27	0.35 BiCMOS
(Dawn et al. 2002)	3.3-4	-133.7	28	0.5 SiGe BiCMOS
(Kampe et al 2005)	1.2-2.5	-118	11.6	0.18 CMOS
(Antoine et al. 2005)	1.848-3.5	-127	-	0.35 SiGe BiCMOS
(Berny et al. 2003)	1.2-1.425	-121	7.2	0.25 CMOS
(Svelto et al. 2000)	1.1-1.45	-126.4	12	0.35 CMOS
(Je-Kwang et al. 2003)	1.59-2.14	-123	32	0.5 SiGe BiCMOS
[this example]	**0.94-1.81**	**-125.5**	**12**	**0.35 CMOS**

Finally, with regard to the tuning range, this work exhibits nearly one octave, only surpassed again by (Kampe et al. 2005). Once again, the 0.18μm CMOS technology has allowed implementing an LC tank with higher quality factor, which permits increasing the tuning range maintaining a low phase noise.

8.5.2 Example 2. High FrequencyVCO

This subsection describes the design of a VCO for the UN-II frequency band up to 5 GHz. In the same way as the DVB-H VCO design has started with the passive elements fabrication and characterization, this example shows its. Previous to the final design, the integrated inductor has been fabricated and characterized. Fig. 8.36 presents a microphotography of the selected inductor.

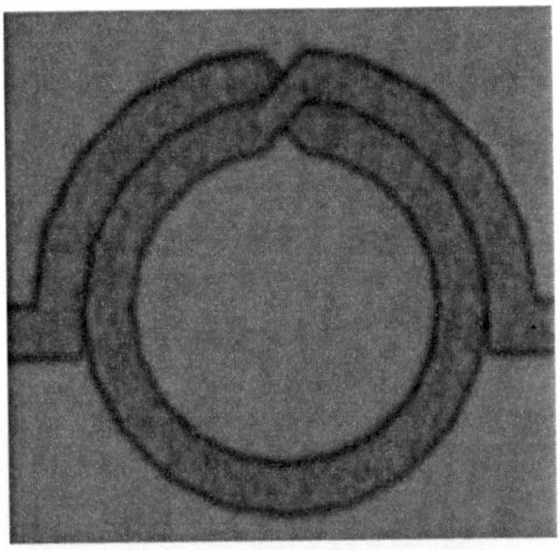

Fig. 8.36 Designed integrated inductor microphotography for the WLAN VCO example

This differential inductor has a value of 0.65 nH and has been optimized with regard to its quality factor (Aguilera and Berenguer 2003), (Q=16 at 5 GHz). This is comparable to the highest values found in the reported works for this working frequency. This passive block has been built in a 20 μm wide thick top metal.

The value of the variable capacitors required for the oscillation in the 5 GHz U-NII band varied from 1.034pF to 1.322pF. These elements have been integrated using the varactors supplied by the UMC CMOS 0.18 μm technology. This range of values had been decided at the simulation stage. The parasitic capacitances of the extracted view of the circuit have lightly corrected the initial theoretical values for these varactors.

The negative resistance amplifier is composed of 2 cross-coupled pairs, one PMOS and one NMOS. The whole core is fed by a tail current source, in a current reuse topology. As it has been introduced at the final stage of the architecture selection, the values of these transistors have to follow a trade off. First, the current source branch needs a higher length transistor to work in a stabilised region. Then, depending on the average value of the resulting current, the values of the MOS pairs need to be high enough to maintain the oscillation, but low enough to assure the objective of minimum power consumption. The high frequency operation demands the lower available length for the core devices. In this case this is 0.18 μm, limited by the technology. Furthermore, the PMOS total width has to be 1.5 times bigger than the NMOS one. This balance has been adopted, among the optimum values, for the appropriate biasing of the output stage connected to the output node.

Finally, fixed capacitors connected to ground have been added to the most critical nodes of the design, in order to diminish the noise contribution of the measurement setup. The voltage supply branch has been connected to a 20 pF distributed capacitor, whereas, the node between the current source and the core has been desensitized by a 7pF capacitor.

8.5.2.1 Layout Considerations

Fig. 8.37 shows the layout view of the VCO designed in the second example. As pointed in the first example, the available chip area within the pads and VCO core has been filled with decoupling capacitances from VDD to ground, but also from the tuning voltage path to ground, in order to decouple the noise coming from DC paths that might contribute to the oscillator's phase noise.

8.5.2.2 VCO Characterization and Results

The die microphotography of the VCO fabricated and a detailed zoom of the core region are shown in Fig. 8.38 and Fig. 8.39 respectively.

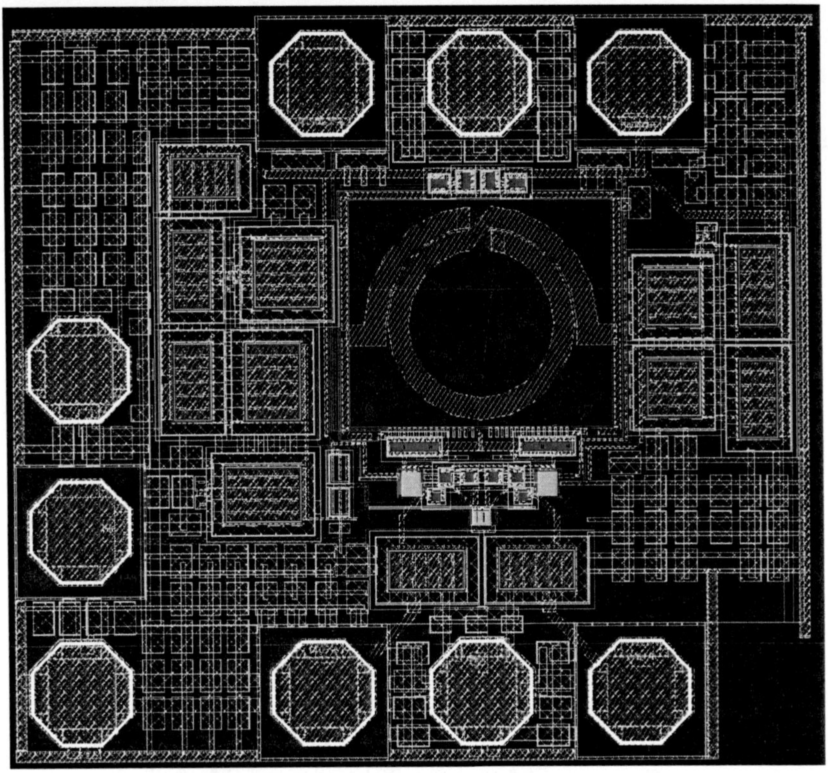

Fig. 8.37 VCO for UN-II frequency band layout

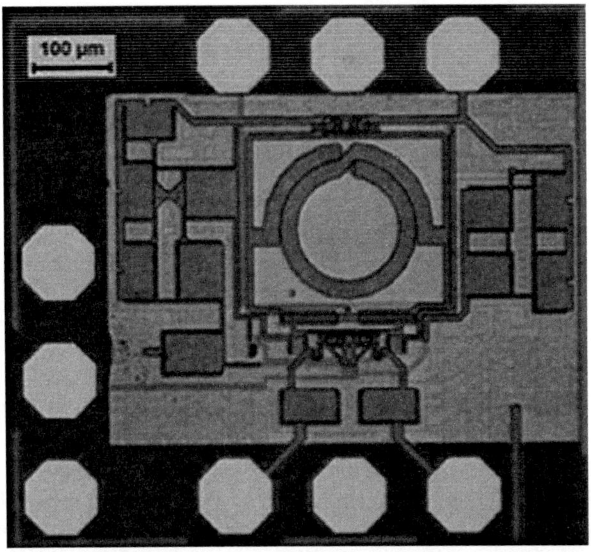

Fig. 8.38 VCO die microphotography for the WLAN VCO example

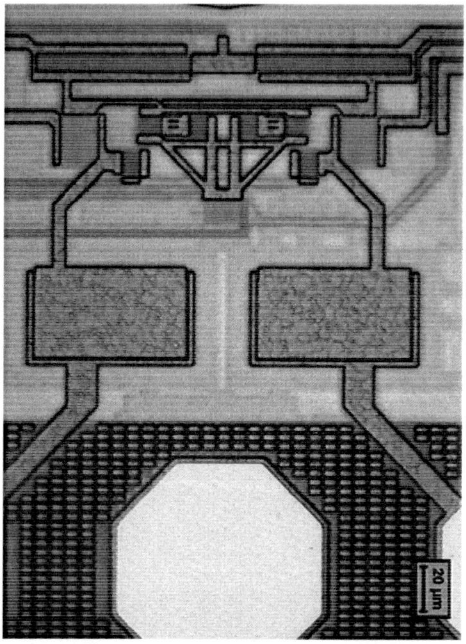

Fig. 8.39 Die microphotography of the core of the VCO for the WLAN VCO example

The characterization of this device has been performed following the measurement setup shown in Fig. 8.40 (different from the setup shown in Fig. 8.32). In this setup, the output tone of the VCO, its power and frequency, is displayed in the spectrum analyzer. A buffer with no influence on the core performance has been employed for the characterization of this block.

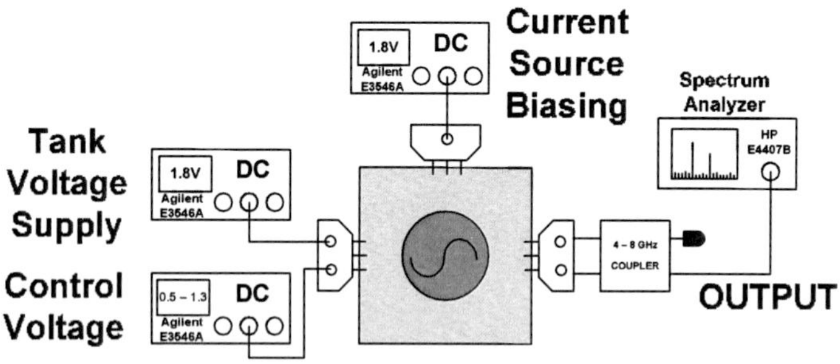

Fig. 8.40 VCO measurement setup for the WLAN VCO example

Fig. 8.41 and Fig. 8.42 present the power level and the frequency of oscillation of the output tone of this VCO depending on the bias current of the core. Both measurements have been obtained for tuning voltages from 0.5 V to 1.3 V.

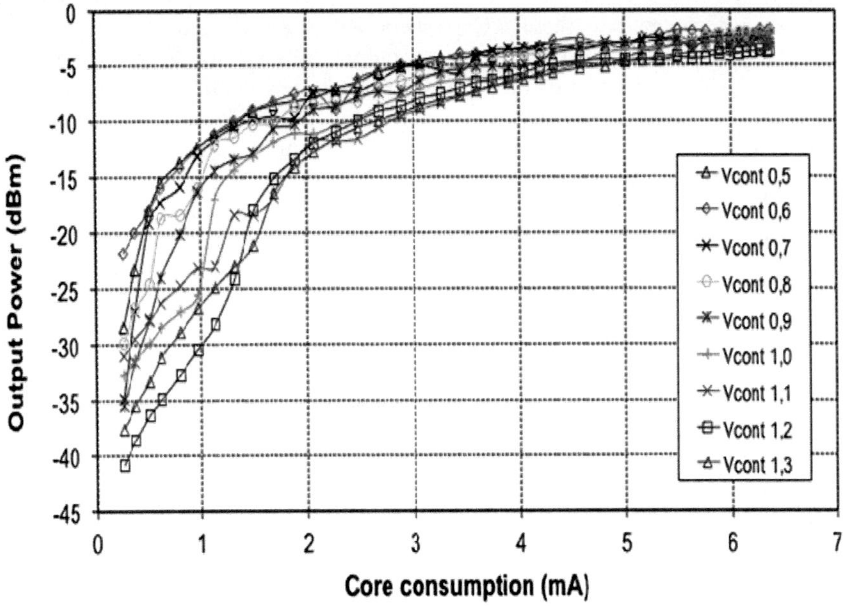

Fig. 8.41 Power measurement Vs Current consumption for the WLAN VCO example

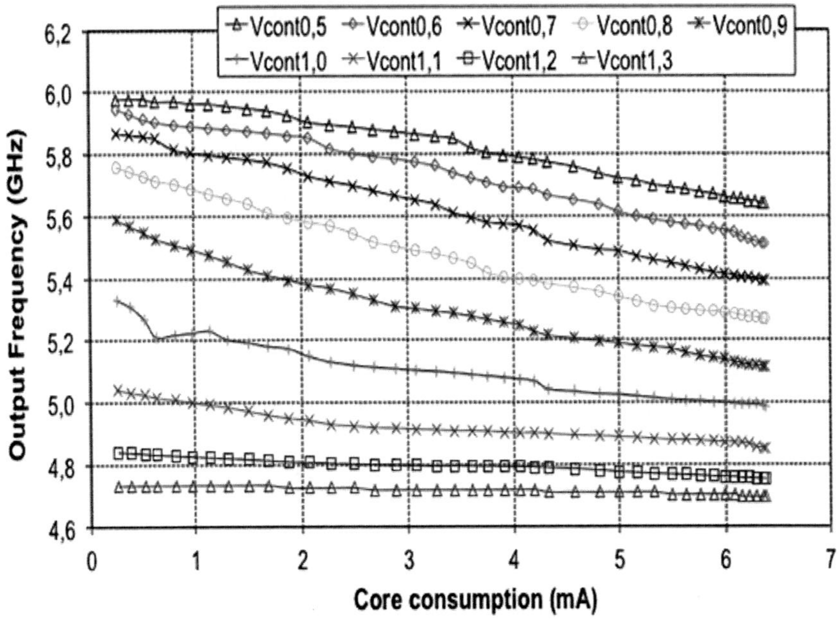

Fig. 8.42 Frequency Vs Current consumption for the WLAN VCO example

This VCO exhibits a phase noise of -111.8dBc/Hz @ 1MHz through the simulation performed to the R&C parasitic extracted view of this device. No measurement can be performed, as the output tone is not locked at a fixed frequency. This simulation result is considered adjusted to the real performance of the VCO, as previous measurements of PLLs with this kind of oscillators and fabricated with the same technology exhibited minimal error percentage from simulations to measurements (Quemada 2006).

Fig. 8.41 and Fig. 8.42 present the Power and the Frequency behavior of this Oscillator depending on the bias current. In the first one, the minimum current available with the application is 1.9mA with a biasing voltage of 1.8V at the gate of the current source transistor. Applying the minimum working voltage for the same current source (0.8V), the threshold current needed to attain the -3dBm output power is 2.8mA.

Above these limits the output power lies largely over the -3dBm line, up to +3dBm, considered the appropriate working range. Fig. 8.42 shows the measured bandwidths obtained from tuning voltage from 0.5 to 1.3V. It is quite interesting note that:

- Lower currents provide higher frequency bandwidths, as it was states in the low power design section.
- Lower currents exhibit higher central frequencies.

This latter result may be explained by the fact that the parasitic capacitors of the MOS transistors are increased by a higher current flowing from their drain source. This effect raises the global capacitor supported by the tank, and as mentioned in Eq. 8.14 and Eq. 8.15, which brings a heightening of the Quality factor Q, which means a narrower oscillation band.

8.5.3 Example 3: High Frequency Divider and Dual-Modulus Prescaler for WLAN (5GHz UNII Band)

The architecture of the frequency divider of this work has been selected accordingly to a frequency plan built for the 5 GHz U-NII band multi-standard applications. The channel bandwidth has been restricted to 20 MHz that is the common application of all the standards under consideration. The multiple channel bandwidths envisaged by the multiple standards considered in the 5 GHz U-NII band, encouraged to select a ΣΔ modulator fractional-N PLL structure. However, the focused blocks in this work have been the high frequency stages that are common to any fractional-N PLL architecture (Razavi 1998). The adoption of the ΣΔ modulator topology would require a minimum redesign of the division ratio control feedback loop.

The complete system set out in this subsection as depicted in Fig. 8.43 is the simulation environment that has demonstrated the capabilities of the High Frequency divider by 2 (HF) and the Dual Modulus Prescaler.

In this case, the frequency plan to tune the 12 center frequencies of the channels depends on the relation of Eq. 8.17: $f_{Channel}$ is the input frequency, DMP_{low} is he lower division coefficient of the Dual Modulus Prescaler, P is the fixed divider after the variable division device, S is the swallow counter coefficient and f_{REF} is the fixed reference of this frequency divider.

$$f_{Channel} = (DMP_{low} \cdot P + S) \cdot f_{REF} \quad \text{Eq. 8.17}$$

With a Dual Modulus Prescaler division ratio of 4/5, the numerical goals for the blocks of the complete frequency divider system are broken down in Table 8.5.

Table 8.5 Frequency plan for the divider for 20 MHz bandwidth channels

Channel centre freq.	Channel Div.	P	S	Output
5180 MHz	2590 MHz	256	12	2.5 MHz
5200 MHz	2600 MHz		16	
5220 MHz	2610 MHz		20	
5240 MHz	2620 MHz		24	
5260 MHz	2630 MHz		28	
5280 MHz	2640 MHz		32	
5300 MHz	2650 MHz		36	
5320 MHz	2660 MHz		40	
5745 MHz	2872.5 MHz		125	
5765 MHz	2882.5 MHz		129	
5785 MHz	2892.5 MHz		133	
5805 MHz	2902.5 MHz		137	

The conclusions extracted from the dividers' bibliography inspection done for this specific example and the low power considerations have determined that for the working frequency and for the technology adopted in that case:

- The first block of the frequency divider has to be a High Frequency divider by 2 (Pellerano et al. 2003).
- A Dual Modulus Prescaler with a division ratio of 4/5 provides the controllable division ratio of this design.

A swallow counter and some basic logic control the DMP. This counter needed to provide 12 control signals from 12 to 137. Thus, an 8-bit counter has been envisaged. The Multiplexor (MUX) and the OR presented in this loop have been necessary for the synchronization of the control. After the variable divider, a fixed low frequency divider of 256 has been selected. This device is composed of 8 low power TSPC FFs in a row. Finally, the last block shown in this figure is a synchronization FF. As outlined by (Pellerano et al. 2003), this re-synchronization avoided output glitches and reduced the phase noise accumulated in the whole divider.

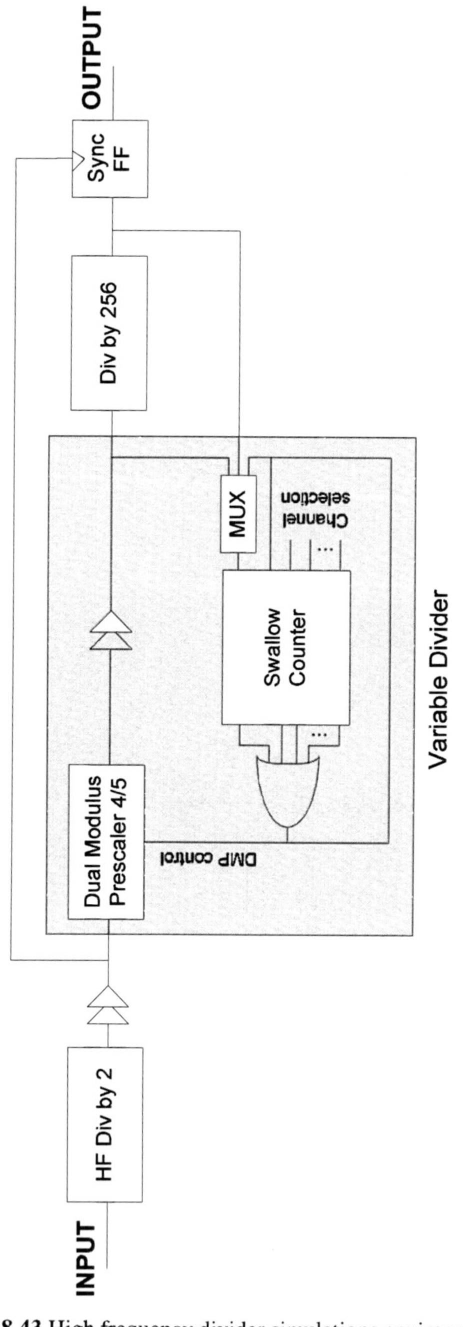

Fig. 8.43 High frequency divider simulations environment

8.5.3.1 Circuit Implementation

The composition of both high frequency dividers, made of TSPC-based FFs is shown in Fig. 8.44.

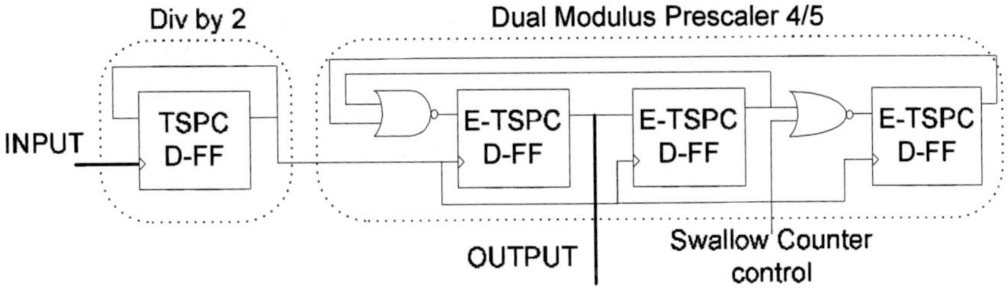

Fig. 8.44 TSPC-based D FF composition of the high frequency dividers

The first stage is the High Frequency divider by 2 and the structure selected for this circuit is the TSPC Flip Flop. The block working at the highest operating frequency is the block that consumes the higher power, thereby in order to minimize this consumption, TSPC is the topology selected. The length of the nine transistors compromised in this circuit design has been tied to the minimum available by the technology (0.18 μm). The widths of the transistors have been studied to provide a better response with the minimum current consumption. Transistors P11, N31 and P13, fromFig. 8.23a, are involved in the output signal definition and trade offs have to be adopted:

- An extremely low value for the width of P13 looses the signal level, whereas a high value holds the pulse and changes the division rate. The final value for W_{P13} is 1.54 μm.
- P11 and N13 form the input inverter of this block. The ratio W_{P11}/W_{N13} defines the pulse width of the signal during the division by 2. The augmentation of this ratio enlarges the pulse width and may also change the division ratio. Finally W_{P11}=1.04 μm and W_{N13}=0.5 μm.
- The remaining 6 transistors of this structure have a 0.5 μm width.
- The layout built for the TSPC based High Frequency divider by 2 explained above is presented in Fig. 8.45, with a dimension area of 15x14 μm^2.

The Dual Modulus Prescaler follows this High Frequency divider by 2. This block receives a control logic signal that is feed-forwarded from several steps beyond. The glitch-free and the high frequency operation capabilities of the E-TSPC make this structure appropriate for this block.

The structure of this block is based on the Dual Modulus Prescaler proposed in (Navarro Soares 1999). Nevertheless, the scaling down of the technology of fabrication has brought the necessity of scaling down all the dimensions of the transistors. The complete schematic of this block is exhibited in Fig. 8.46.

Fig. 8.45 Layout of the TSPC Flip Flop

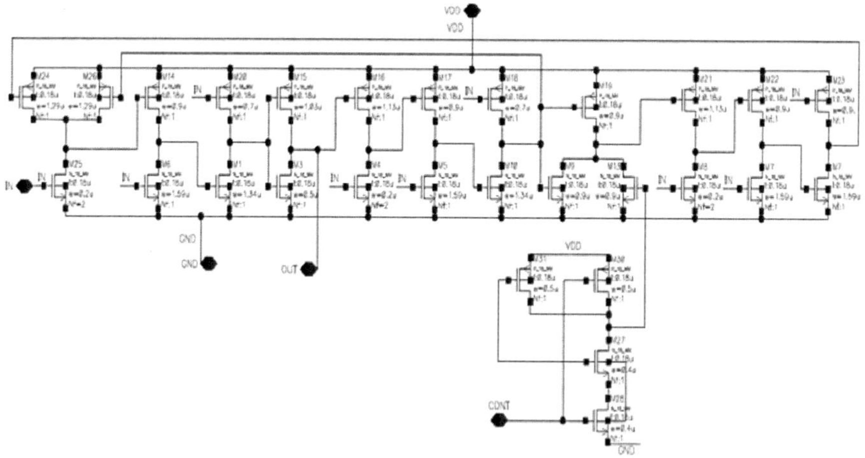

Fig. 8.46 Dual Modulus Prescaler schematic

8.5.3.2 Simulation Results

The high frequency dividers have been simulated with the defined environment in order to test their running capabilities. The simulation of the joint operation of the two first blocks is exhibited in Fig. 8.47, at a frequency of 5.25 GHz. In this case a perfect division by 8 is performed.

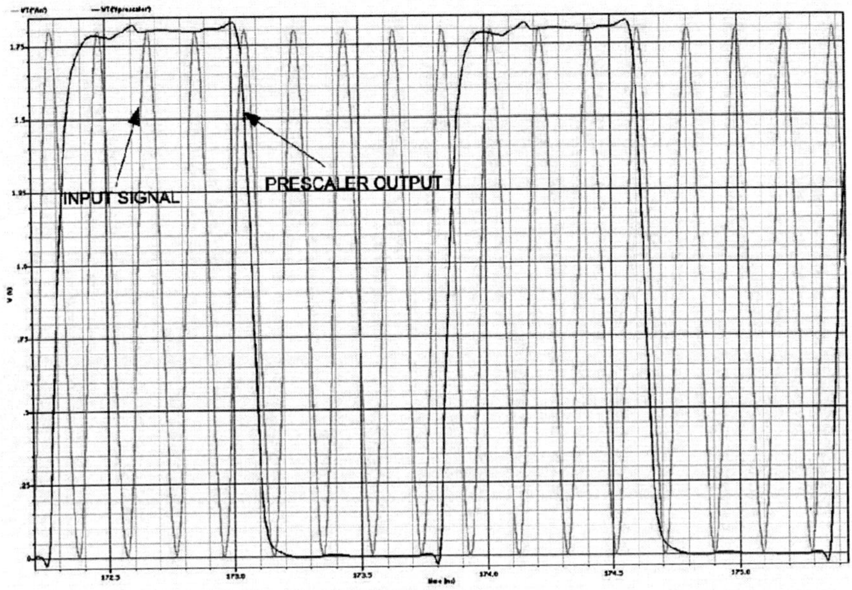

Fig. 8.47 TSPC-based divider by 2 and Prescaler simulation

The simulation of the last stages of the divider, shown in Fig. 8.43, and the control signals are presented in Fig. 8.48. In this graph, the 2.5 MHz output is extracted from a 5.25 GHz input signal.

Table 8.6 details the current consumption of each block for the centre of the two main frequency region (5 GHz U-NII 1&2 and 5 GHz U-NII 3). These results have been deduced from the average transient response of the current consumed by each device during 800 ns (2 output cycle and about 4400 input cycles.)

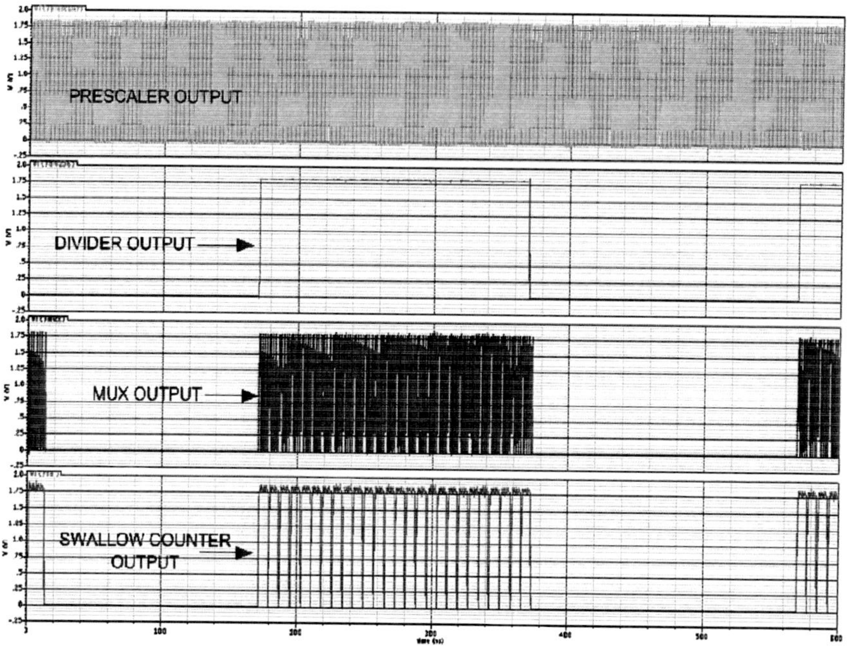

Fig. 8.48 Divider output and control signals

Table 8.6 Frequency divider blocks current consumption

Block	Current Consumption 5.25 GHz	Current Consumption 5.775 GHz
HF divider by 2	205.3 μA	212.7 μA
4/5 Prescaler	381.9 μA	359.3 μA
Divider by 256	35.68 μA	38.39 μA
Sync FF	72.21 μA	41.65 μA
Swallow Counter	47.02 μA	59.02 μA
MUX and control logic	19.64 μA	28.92 μA
TOTAL	761.75 μA	739.98 μA

References

(Adin 2007) Adín, I.: RF CMOS ICs Design applied to Multistandard Wireless Applications for the 5 GHz U-NII band. PhD. Thesis (2007)

(Aguilera and Berenguer 2003) Aguilera, J., Berenguer, R.: Design and Test of Integrated Inductors for RF applications. Kluwer Academic Publishers, Dordrecht (2003)

(Ahola et al. 2004) Ahola, R., et al.: A Single-Chip CMOS Transceiver for 802.11a/b/g Wireless LANs. IEEE Journal of Solid-State Circuits 39(12), 2250–2258 (2004)

(Antoine et al. 2005) Antoine, P., et al.: A direct-conversion receiver for DVB-H. IEEE Journal of Solid-State Circuits 40, 25–36 (2005)

(Behzad et al. 2003) Behzad, A.R., et al.: A 5-GHz Direct-Conversion CMOS Transceiver Utilizing Automatic Frequency Control for the IEEE 802.11a Wireless LAN Standard. IEEE Journal of Solid-State Circuits 38(12), 2209–2220 (2003)

(Berny et al. 2003) Berny, A.D., et al.: A wideband low-phase-noise CMOS VCO. In: Proceedings of the IEEE 2003 Custom Integrated Circuits Conference, September 21-24, pp. 555–558 (2003)

(Craninckx and Steyaert 1998) Craninckx, J., Steyaert, M.: Wireless CMOS Frequency Synthesizer Design. Kluwer Academic Publishers, Dordrecht (1998)

(Dawn et al. 2002) Wang, D., et al.: A fully integrated GSM/DCS/PCS Rx VCO with fast switching auto-band selection. In: IEEE Radio and Wireless Conference, RAWCON 2002, August 11-14, pp. 209–212 (2002)

(Egan 2000) Egan, W.F.: Frequency Synthesis by Phase Lock. Wiley Interscience, Hoboken (2000)

(Hajimiri and Lee 1998) Hajimiri, A., Lee, T.H.: A General Theory of Phase Noise in Electrical Oscillators. IEEE Journal of Solid-State Circuits 33(2), 179–194 (1998)

(Hajimiri and Lee 1999) Hajimiri, A., Lee, T.H.: Design Issues in CMOS Differential LC Oscillators. IEEE JSSC 34(5) (May 1999)

(Herzel et al. 2003) Herzel, F., et al.: An Integrated CMOS RF Synthesizer for 802.11a Wireless LAN. IEEE Journal of Solid-State Circuits 38(10), 1767–1770 (2003)

(Huang 2001) Huang, Q.: Circuit Design for Wireless Communications. Curso de RF, Lausanne, Suiza (June 2001)

(Je-Kwang et al. 2003) Je-Kwang, C., et al.: A 2-GHz wide band low phase noise voltage-controlled oscillator with on-chip LC tank. Presented at Proceedings of the IEEE Custom Integrated Circuits Conference (2003)

(Kampe et al. 2005) Kampe, A., et al.: An LC-VCO with one octave tuning range. Presented at Proceedings of the 2005 European Conference on Circuit Theory and Design (2005)

(Lee and Hajimiri 2000) Lee, T.H., Hajimiri, A.: Oscillator Phase Noise: A Tutorial. IEEE Journal of Solid-State Circuits 34(3), 326–336 (2000)

(Leeson 1966) Leeson, D.B.: A Simple Model of Feedback Oscillator Noise Spectrum. Proc. IEEE L. 54, 329–330 (1966)

(Leung and Luong 2004) Leung, G.C.T., Luong, H.C.: A 1-V 5.2-GHz CMOS Synthesizer for WLAN Applications. IEEE Journal of Solid-State Circuits 39(11), 1873–1882 (2004)

(Levantino et al. 2004) Levantino, S., et al.: Phase Noise in Digital Frequency Dividers. IEEE Journal of Solid-State Circuits 39(5), 775–783 (2004)

(Manetakis et al. 2004) Manetakis, K., et al.: A CMOS VCO with 48% tuning range for modern broadband systems. Presented at Proceedings of the IEEE Custom Integrated Circuits Conference (2004)

(Mendizabal 2006) Mendizábal, J.: Design of a highly integrated dual RF front-end for GPS and Galileo in 0.35μm SiGe process, PhD Thesis (2006)

(Navarro Soares 1999) Soares, N., Noije, V.: A 1.6-GHz dual modulus prescaler using the extended true-single-phase-clock CMOS circuit technique (E-TSPC). IEEE Journal of Solid-State Circuits 34(1) (1999)

(Pellerano et al. 2003) Pellerano, S., et al.: 13.5-mW, 5-GHz WLAN, CMOS frequency synthesizer using a true single phase clock divider. In: VLSI Circuits 2003 (2003)

(Perrott 2003) Perrott, M.: High Speed Communication Circuits and Systems: High Speed Frequency Dividers. Mitopencourseware, Massachusetts Institute of Technology, USA (2003)

(Quemada 2006) Quemada, C.: Metodología de diseño de PLLs aplicada al desarrollo de un sintetizador de frecuencia integrado CMOS para el estándar de WLAN IEEE 802.11[a], PhD Thesis (2006)

(Quemada et al. 2009) Quemada, C., et al.: Design Methodologies for RF CMOS Phase Locked Loops. Artech House Publishers, Boston (2009)

(Rategh et al. 2000) Rategh, H., et al.: A CMOS Frequency Synthesizer with an Injection-Locked Frequency Divider for a 5-GHz Wireless LAN Receiver. IEEE Journal of Solid-State Circuits 35(5), 780–787 (2000)

(Razavi et al. 1995) Razavi, B., et al.: Design of High-Speed, Low-Power Frequency Dividers and Phase-Locked Loops in Deep Submicron CMOS. IEEE Journal of Solid-State Circuits 30(2), 101–109 (1995)

(Razavi 1996) Razavi, B.: A Study of Phase Noise in CMOS Oscillators. IEEE Journal of Solid-State Circuits 31(3), 331–343 (1996)

(Razavi 1998) Razavi, B.: RF Microelectronics. Prentice-Hall, Englewood Cliffs (1998)

(Razavi and Lam 2000) Razavi, B., Lam, C.: A 2.6-GHz/5.2-GHz Frequency Synthesizer in 0.4-μm CMOS Technology. IEEE Journal of Solid-State Circuits 35(5), 788–794 (2000)

(Rohde and Newkirk 2000) Rohde, U.L., Newkirk, D.P.: RF/Microwave Circuit Design for Wireless Applications. Wiley Interscience, Hoboken (2000)

(Sjoland and Sjoblom 2007) Sjoland, H., Sjoblom, P.: Measured CMOS Switched High-Quality Capacitors in a Reconfigurable Matching Network. IEEE Transactions on Circuits and Systems II: Express Briefs 10, 858–862 (2007)

(Svelto et al. 2000) Svelto, F., et al.: A 1.3 GHz low-phase noise fully tunable CMOS LC VCO. IEEE Journal of Solid-State Circuits 35, 356 (2000)

(Valla et al. 2005) Valla, M., et al.: A 72-mW CMOS 802.11a Direct Conversion Front-End With 3.5-dB NF and 200-KHz 1/f Noise Corner. IEEE Journal of Solid-State Circuits 40(4), 970–977 (2005)

(Vassiliou et al. 2003) Vassiliou, I., et al.: A Single-Chip Digitally Calibrated 5.15-5.825-GHz 0.18-μm CMOS Transceiver for 802.11a Wireless LAN. IEEE Journal of Solid-State Circuits 38(12), 2221–2229 (2003)

(Wang 2003) Wang: A 1.8V 3mW 16.8GHz Frequency Divider in 0.25μm CMOS. In: IEEE International Solid-State Circuits Conference, pp. 196–197 (2000)

(Wohlmuth and Kehrer 2002) Wohlmuth, H., Kehrer, D.: A High Sensitivity Static 2:1 Frequency Divider up to 27 GHz in 120nm CMOS. In: Proc. Eur. Solid State Circuits Conf. (ESSCIRC), pp. 823–826 (2002)

(Yu et al. 2006) Yu, D., et al.: Design and Optimization of the Extended True Single-Phase Clock-Based Prescaler. IEEE Transactions on Microwave Theory and Techniques 54(11) (2006)

(Yuan and Svensson 1989) Yuan, J., Svensson, C.: High-Speed CMOS Circuits Technique. IEEE Journal of Solid-State Circuits 24(1), 62–70 (1989)

(Zargari et al. 2002) Zargari, M., et al.: A 5-GHz CMOS Transceiver for IEEE 802.11a Wireless LAN Systems. IEEE Journal of Solid-State Circuits 37(12), 1688–1694 (2002)

(Zhang et al. 2003) Zhang, P., et al.: A 5-GHz Direct-Conversion CMOS Transceiver. IEEE Journal of Solid-State Circuits 38(12), 2232–2237 (2003)